T0174456

CAMBRIDGE COUNTY GEOGRAPHIES

General Editor: F. H. H. GUILLEMARD, M.A., M.D.

WEST LONDON

Cambridge County Geographies

WEST LONDON

by

G. F. BOSWORTH, F.R.G.S.

With Maps, Diagrams and Illustrations

Cambridge :

at the University Press

1912

CAMBRIDGE UNIVERSITY PRESS
Cambridge, New York, Melbourne, Madrid, Cape Town,
Singapore, São Paulo, Delhi, Mexico City

Cambridge University Press
The Edinburgh Building, Cambridge CB2 8RU, UK

Published in the United States of America by Cambridge University Press, New York

www.cambridge.org
Information on this title: www.cambridge.org/9781107663602

First published 1912
First paperback edition 2013

A catalogue record for this publication is available from the British Library

ISBN 978-1-107-66360-2 Paperback

CONTENTS

a 3

ILLUSTRATIONS

MAPS

The illustrations on pp. 44, 49, 131, and 192 are reproduced from *Literary London* by courtesy of Mr T. Werner Laurie; that on p. 82 is from a photo kindly supplied by Sir W. Christie; that on p. 93 from a photo supplied by Messrs Doulton and Co., Ltd; the design on p. 175 is reproduced by kind permission of *The Building News*; thanks are also due to the Bishop of London for the photo on p. 158 and to Mr S. Bewsher, Bursar of St Paul's School, for that on p. 226. The portraits on pp. 231, 233, 240, 241, 242, 245, 247 and 248 are from photographs by Mr Emery Walker; that on p. 237 is reproduced by kind permission of Pembroke College, Cambridge.

1. County and Shire. The County of London. The word *London*: its Origin and Meaning.

The main divisions of our country are known as counties, and, in some instances, as shires. When the word shire is used, it is added to the county name. For instance, we speak of the county of Kent, or of the county of Bedford; but while the word shire is not added to the name of Kent, it may be to that of Bedford. Thus we write the county of Bedford, or Bedfordshire, but not the county of Bedfordshire. Such an expression would be wrong and superfluous, for the word shire is now practically equivalent to the later word county.

Although, however, we now call all the divisions of England and Wales counties, that title is not historically accurate. Some counties, such as Kent, Essex, and Sussex, are really survivals of various old English kingdoms, and for more than a thousand years there has been but little alteration either in their boundaries or their names.

The divisions now known as Bedfordshire, Hertfordshire, and Wiltshire are so called because they were *shares* or portions cut off from larger kingdoms. Thus Bedford-shire and Hertfordshire were shares or portions of a very large kingdom known as Mercia, while Wiltshire was a share or portion of Wessex. It is not necessary to enlarge further on this distinction, but it is well to have a correct idea of the origin of our counties. For many years it was wrongly stated that Alfred divided England into counties. The statement is incorrect, for we know that some of the counties were in existence before his time, while others were formed after his death.

It may be stated here that the object of thus dividing our country into counties was partly military and partly financial. Every shire had to provide a certain number of armed men to fight the king's battles, and also to pay a certain proportion of the king's income. In each case a "shire-reeve," or sheriff as we now call him, was appointed by the king to see that the shire did its duty in both respects. After the Norman Conquest, the government of each shire was handed over to a count, and from that time these divisions have been called counties.

In England the divisions or ancient counties numbered forty until the year 1888. Then it was decided to form the Administrative County of London, under the pro-visions of the Local Government Act of that year. It is to be noted that, although London is the latest of the forty-one counties, it is not known as an "ancient" county, for it was constituted an administrative area from parts of the ancient counties of Middlesex, Surrey, and Kent. Thus it comes about that London, the capital of the British Empire, the greatest city in the world, and once the capital of the county of Middlesex, is now an Administrative county.

There is another London, which is often called "Greater London," but with that we do not propose to deal, as that enlarged area takes in many parishes and districts that are outside the boundaries of the administrative county, and extend into Hertfordshire and Essex.

Now with regard to the name *London*, there is great diversity of opinion as to its origin and meaning. We shall not, however, be wrong if, in giving some of the opinions on this subject, we state that the earliest historic monument of London is its name. The word *Londinium* first appears in Tacitus under the year A.D. 61 as that of an *oppidum* not dignified with the name of a colony, but celebrated for the gathering of dealers and commodities.

It follows from this early notice that *Londinium* must have been founded long before A.D. 61, and historians have come to the conclusion that the Roman *oppidum* was built on the site of an earlier Celtic village, and that the name *Londinium* is the Latinised form of *Llyn-Din*, i.e. the lake-fort.

Some writers have endeavoured to explain the name from other Welsh roots, but nothing is so uncertain as the origin of some place-names. Geoffrey of Monmouth thinks that London was called *Caer-Lud* after a King Lud of Celtic history, and even some recent writers have come round to this view and say that London means Lud's-town. This last derivation may be mere conjecture, although it is in harmony with tradition.

A View of LONDON as it appeared before the dreadful Fire A.D. 1666

References

1 St Paulo	7 St Sepulchres
2 St Dunstans	8 Bow Church
3 Temple	9 Guild hall
4 St Brides	10 St Michaels
5 St Andrew	11 St Lawrence Poultney
6 Baynards Castle	12 Old Swan

13 London Bridge	20 St Mary overs
14 St Magnus Parish	21 Winchester house
15 Billingsgate	22 The Globe
16 Custom house	23 The Bear Garden
17 Tower	24 Hampsted
18 Dr Wharf	25 Highgate
19 St Olaves	26 Hackney

It may be mentioned that Geoffrey of Monmouth wrote early in the twelfth century, and gives a legend of the founding of London. This describes how Brutus came over from Troy and formed the plan of building a city. When he came to the Thames he found a site on its banks suitable for his purpose. There he built a city, calling it *Troia Nova*, i.e. New Troy, which was afterwards corrupted into *Trinovantum*. As time passed on, King Lud built walls and towers round the city; and when he died, his body was buried by the gate which is called in the Celtic speech " Porthlud," but in the Saxon " Ludesgata "—our Ludgate.

Here then we have the legend of the origin of London in pre-Roman days, and it may be founded on some genuine folk-stories of Celtic origin. At any rate, it explains the fact that the Roman attempt to change the name to *Augusta* completely failed, for the early name Llyn-din, or Caer-Lud, held its own in the affections of the Britons. Whatever conclusion we reach with regard to the origin of the name London, we feel sure that it was a village of some importance before the Roman occupation, as prehistoric and early relics are often found on the site.

Thus it comes about that London has almost an unbroken record extending over 2000 years, and whether as Llyn-din, or Augusta, or Londinium, or London, occupies a commanding place in our country's history.

2. General Characteristics. Position and Natural Conditions. Why London is our Capital.

There may be doubts as to the origin of London and the exact meaning of its name, but there can be no doubt as to its two thousand years of unbroken history and that it exerts a great fascination over the imagination of Englishmen. It has been well remarked that "London has a charm all her own; it is that of a history as romantic and as interesting to Englishmen as that of Ancient Rome was to the Romans. As Ancient Rome once was, so is London now the centre of civilisation."

In this chapter we shall first glance at some of the general characteristics of London, and then pass on to consider its position, and why it came to be chosen as our capital. There are people who would argue that London is a most unsuitable site for a capital, but we have to remember that it has stood the great practical test of centuries and has won its way to the foremost place against the competition of other cities that were officially favoured. Thus York was the chief Roman centre of administration, and Winchester was the chief town of Wessex and became the capital when the kings of Wessex were supreme over all England.

It is sometimes easy to give the characteristics of a city or of some place of historic interest. But in dealing with London we have to think of at least two cities, round which have grown numerous towns that would each be considered large in the provinces. The immensity of London is so overwhelming, and its variety is so amazing, that we are not surprised to find how differently London is characterised by poets and historians.

Wordsworth was charmed with the sight of London

The Embankment looking Citywards from Charing Cross

from Westminster Bridge, and in one of his sonnets exclaims :—

> "Earth has not anything to show more fair;
> Dull would he be of soul who could pass by
> A sight so touching in its majesty";

Byron looked upon it as "A mighty mass of brick, and smoke, and shipping." A French writer calls it "a province in brick"; and one of our own literary men characterises it as "a squalid village." Heine, the great German writer, gives his idea of London as "a forest of houses, between which ebbs and flows a stream of human faces, with all their varied passions—an awful rush of love, hunger, and hate."

There is some truth in each of these various attempts to give an idea of London, but of course they are all short of leaving the correct impression. Probably no one man is capable of giving a true picture of London, for there are so many aspects of the modern city. Its immense population and the strange variety of races are sure to have their effect on one class of observers. Others will be struck by the contrasts between the princely palaces of the rich and the filthy hovels of the poor, or between the magnificent squares and the squalid slums. In no other city in the world is there such a striking difference between historic buildings which date from the Conquest and the modern structures of stone and marble which have supplanted the wooden houses of the Stuart period.

Such, then, are a few of the most remarkable characteristics of London as it is to-day. It is not possible to deal further with this subject in the present book, so we will proceed to consider the position of London and what effect the choice of the site of the City by the early founders has had on its subsequent prosperity.

It will be well to look at an early map of the capital showing the marshes on either side of the Thames. We shall then get some idea of what the Thames was like in British days. Then, the river must have looked like a broad lake with here and there a small island rising out of the water. When the tide was high, the river was converted into an arm of the sea, while at low water it was a vast marsh through which the stream wound its way in irregular fashion. It has been estimated that at

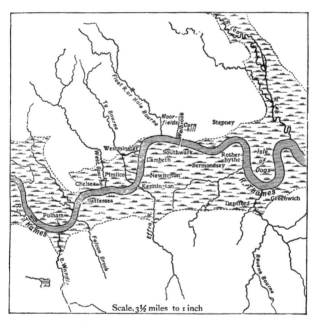

Ancient London and its Surrounding Marshes

least half of modern London is built on this marsh, which extended from Fulham on the west to Greenwich on the east.

In those far-off days the marsh was the resort of wild duck, wild geese, herons, and other water birds flying

over it in myriads.   Altogether we can picture the site
of London two thousand years ago as a dreary and desolate
place, and one of the first questions that arises from this
knowledge is, How came London to be founded on a
marsh ?

There are many reasons why London was founded
on the present site, and if we consider a few of them it
will help us to understand its growth and development.
Of course we are referring to the site of London as it was
in the time of the British founders, and at the period of
the Roman Conquest.   The evidence goes to show that
the earliest centre of the City was on the east side of the
Walbrook at the head of London Bridge.   Now taking
that district as the nucleus of the early city, we find that
London was built on the first place going up the river
where any tract of dry land touched the stream.   We
also find that it is a tract of good gravel soil, well supplied
with water, and not liable to flooding.   These were most
important considerations in selecting the site of a city in
those early days, just as they are at the present time.

It will be seen that this area of good land was chosen
on the river Thames, so that the waterway was a means
of defence, and a highway which could be traversed both
up and down by means of the British boats.   The site
was not very near the sea, and that fact was also an
advantage, for the small boats of the Britons could not
venture on the waves of the Lower Thames.   There is
no doubt that the place was founded on a site about
60 miles from the coast, because it was not open to attack
from the enemies who came over the sea.   It is here

worth mentioning that *London* and *Thames* are both Celtic words, and are the only names remaining in this area to remind us of the British occupation.

There is one other reason we may consider in this connection. London was placed on a tidal river, and thus it carried boats laden with merchandise or passengers far up the river to the west, and far down the river to the east. We may be sure that the Britons made use of the tide, and the Romans, who had been accustomed to the nearly tideless waters of the Mediterranean, soon learnt the value of the ebbing and flowing of the Thames.

Thus we may conclude that the earliest site of London was on land about 50 feet above the level of the tide, and the position was admirably adapted for defence, for it was almost impregnable. Green, in *The Making of England*, remarks that London "sheltered to east and south by the lagoons of the Lea and the Thames, guarded to the westward by the deep cleft of the Fleet, saw stretching along its northern border the broad fen whose name has survived in our modern Moorgate....The 'dun' was in fact the centre of a vast wilderness. Beyond the marshes to the east lay the forest tract of southern Essex. Across the lagoon to the south rose the woodlands of Sydenham and Forest Hill, themselves but advance guards of the fastnesses of the Weald. To the north the heights of Highgate and Hampstead were crowned with forest masses, through which the boar and the wild-ox wandered without fear of man to the days of the Plantagenets. Even the open country to the west was but a waste. It seems to have formed the border-land between two

British tribes who dwelt in Hertford and in Essex—its
barren clays were given over to solitude by the usages of
primeval war."

Besides the geographical reasons that account for the
greatness of London, there are also historical and political
reasons for its prosperity and development.   Bristol and
Liverpool on the west, and Plymouth and Southampton
on the south, are equally well placed, and have enjoyed
exceptional facilities for the cultivation of foreign trade.
But while these and other towns have been fettered by
the action of their feudal lords, London has had no overlord but the king.   The City has always had rule over its
own district, and was not controlled by any outside power.
Thus it comes about that London has distanced all rivals,
such as York and Winchester, and now stands without a
peer, the capital of the British Empire and the greatest
city of the world.

3.   Size.   Boundaries.   Development. History of Growth.   London of the Romans, of the Saxons, of the Normans.   Medieval London. Stuart London.

As we have already seen in a former chapter, England
was formerly divided into 40 geographical counties, but in
1888 it was decided to form the Administrative County
of London.   The number of Geographical Counties is now
41; but England is also divided by the Local Government

Act of 1888 into 50 Administrative Counties. Some of the larger counties were then divided into two or more portions, so that the old idea of 40 counties has become obsolete, and we now speak not only of Sussex and Suffolk, but also of East Sussex and West Sussex, of East Suffolk and West Suffolk. It is well to make this point quite clear, so that we may understand London's position as a county.

Of the 41 geographical counties in England, London is the most recently formed, it is the most important, and it is the smallest in point of size. A reference to the diagrams at the end of the book will illustrate its area compared with that of England and Wales. London contains 74,839 acres or 116·9 square miles, and is thus about $\frac{1}{498}$ of England and Wales. The heart of the county is called the City of London, and this is about one square mile in area.

A glance at the map of the County of London will show that it is an irregularly-shaped area divided into two unequal parts by the many windings of the river Thames: the northern portion is entirely formed from Middlesex, while the southern portion has been taken from both Surrey and Kent. The northern portion contains about two-fifths of the entire area, but it is in many respects the more important of the two divisions.

The length of the county measured from Hammersmith on the west to Plumstead on the east is about 17 miles, while the breadth from Holloway in the north to Streatham in the south is about 11 miles. It will be noticed that there is a small portion of the county on the

Essex side of the Thames. This is known as North Woolwich, and before 1888 this district was part of Kent although it is actually in the county of Essex.

Except on the east side, where it is bounded for some miles by the river Lea, the boundaries are not physical. Middlesex forms the boundary on the north and partly on the west, while Surrey bounds it partly on the west and south, and Kent partly on the south and east.

Before we go further, it will be well to understand that the present volume on the western portion of London includes all the district west of the boroughs of Islington, Finsbury, City of London, Southwark, and Camberwell, and has an area of 33,070 acres. This western portion of London comprises 13 out of the 29 boroughs into which the county is divided. Ten boroughs in the western portion are on the north of the Thames, and the remaining three lie south of that river. The southern portion is much larger than the northern portion, although it is not so important, for we must always remember that, for many centuries, London as a city was only built on the north bank of the Thames.

The line of division that is chosen for this volume is purely arbitrary, and is merely for purposes of convenience. In the eastern portion we get the City of London with its surrounding boroughs, and in the western portion we have the City of Westminster and its neighbouring boroughs. Lewisham in the western portion, and Woolwich in the eastern portion, are the largest boroughs; and Holborn in the western, and Finsbury in the eastern, are the smallest.

Having given these facts and figures relating to the size of the present County of London, we may briefly glance at a little history as to its growth and development from the earliest times.   It would be quite impossible within the limits of this book to go into details ; but we can give a few ideas as to its size and condition at three or four turning points in its history.

In British times we must fall back on conjecture, but we have also the aid of geography and geology.   The foundations of the facts that prove the condition of the earliest London are the waste, marshy ground, with little hills rising from the plains, and the dense forest to the north.   The position of the town on the Thames proves the wisdom of those who chose the site, although the frequent overflowing of the river must have hindered its progress.

Under the Romans, the city became the chief residence of merchants and the great mart of trade.   The Romans probably built a fort where the Tower now stands, and afterwards the walls surrounding the town were erected. Then Londinium took its proper place among the Roman cities of Britain, for it was on the high road to York and the starting-point of most of the Roman roads in Britain. The two chief events in the history of Roman London are the building of the bridge and the building of the wall.   The exact date of the building of the wall cannot be given, but we know that in 350 A.D. it did not exist, while in 368 A.D. the town with its villas, its gardens, and its township was enclosed.   A reference to the map will show the circuit of the wall, with its gates and

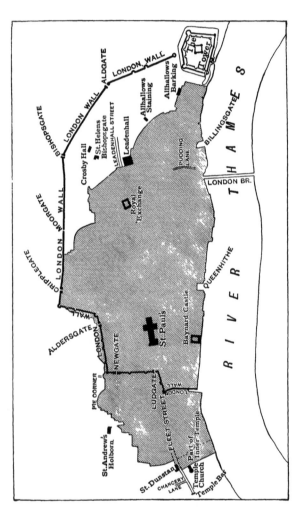

Plan of Old London: showing the Wall and Gates

(The shaded area was that destroyed by the Great Fire)

forts.   London within the wall occupied an area of about
380 acres, and was about $3\frac{1}{4}$ miles in circumference.   This
Roman wall round London was of the utmost importance
in the history of the city, and even to this day it forms in
part the City boundary.

When the Roman legions left Britain, London had a
very mixed population of traders.   The inhabitants were
defenceless and at the mercy of the invader.   The Saxons
conquered the eastern portion of England, and named it
Essex.   London became the capital of the East-Saxon
kingdom.   Saxon London was a wooden city, surrounded
by walls, which probably marked the same enclosure as
the Roman city.   In the seventh century the city had
become a prosperous place, and was peopled by merchants
of many nations.   It was a free trading town, and was
also the great mart of slaves.   In the eighth and ninth
centuries it was frequently harried and laid waste by the
Danes, but the great turning-point in its history was in
886 A.D., when King Alfred restored it and introduced a
garrison of men for its defence.   From this year to the
present time London has been in the front rank of our
cities, and at the Norman Conquest it became, without
a rival, the capital of England.   The further growth
and development of the city were now very marked, and
William I granted a charter to William the Bishop, and
Gosfrith the Portreeve, who is supposed to be Geoffrey
de Mandeville.

If we want to get further particulars of the growth of
London, we must refer to the literature of the fourteenth
and subsequent centuries.   London places are frequently

mentioned in *Piers Plowman*; while Hoccleve, Gower, Lydgate, and Chaucer are invaluable to the student of early London life. The *London Lickpenny*, a work often attributed to Lydgate, is a valuable record of London

A Party of Pilgrims
(*From a MS in the British Museum*)

life at the end of the fourteenth century. In it are related the adventures of a poor Kentish man who went to London in search of justice, but could not find it for lack of money. Chaucer gives us many pictures of the London of his day, and the portraits of the pilgrims in the Prologue to the *Canterbury Tales* show us the men and women who were to be seen daily in the streets of London.

When we come down to the Stuart period, we find that London had about 150,000 people in the reign of James I, and in the reign of Charles II we are told that "the trade and very City of London removes westward, and the walled City is but one-fifth of the whole pile." Lord Macaulay made a special study of the state of London in 1685, and the following extract from his *History of England* gives a very picturesque account of the condition of the City more than two hundred years ago. He writes thus :—"Whoever examines the maps of London which were published towards the close of the reign of Charles the Second will see that only the nucleus of the present capital then existed. The town did not, as now, fade by imperceptible degrees into the country. No long avenues of villas, embowered in lilacs and laburnums, extended from the great centre of wealth and civilisation almost to the boundaries of Middlesex, and far into the heart of Kent and Surrey. In the east no part of the immense line of warehouses and artificial lakes which now stretches from the Tower to Blackheath had been projected. On the west, scarcely one of those stately piles of building which are inhabited by the noble and wealthy was in existence ; and Chelsea...was a quiet country village with about a thousand inhabitants. On the north cattle fed, and sportsmen wandered with dogs and guns over the site of the borough of Marylebone, and over far the greater part of the space now covered by... Finsbury and the Tower Hamlets. Islington was almost a solitude ; and poets loved to contrast its silence and repose with the din and turmoil of the monster London. On the south the capital is now connected with its suburb by several bridges, not inferior in magnificence and solidity to the noblest works of the Caesars.   In 1685, a single line of irregular arches, overhung by piles of mean and crazy houses, and garnished, after a fashion worthy of the naked barbarians of Dahomey, with scores of mouldering heads, impeded the navigation of the river."

Lord Macaulay wrote this interesting sketch of Stuart London more than 60 years ago, when the population of the metropolis was under two millions. Since Macaulay's time London has increased enormously both in area and population, and the contrast between the early Victorian London and that of to-day is almost as striking as that drawn by the great Whig historian. Although a term has been put on its extent by the Act of 1888, its population has increased and, as we shall see in subsequent chapters, its development in trade and commerce is also progressive.

4.   London Parks, Commons, and Open Spaces in the N.W. and S.W.

If we look at any map of London showing the parks, commons, and open spaces within its boundaries we shall at once realise that Londoners are very fortunate in being so well provided with municipal "lungs." The first idea of many people who do not know London is that the

Metropolis is nothing more than a wilderness of brick and mortar. This, we shall find, is far from being true; and probably no other capital in the world has such extensive breathing spaces for its people as ours. The finest and largest parks are, as we might expect, in the western portion of the county; but we must remember that the people in the north-east have Epping Forest, which, although in Essex, is yet maintained by the City Corporation and is known as London's Playground.

Now first we will endeavour to get a good idea of the extent of London's parks and open spaces; then we will consider some of their characteristics; and finally we will pass in review those that are situated in the western portion of the County of London.

The parks, commons, and open spaces within the County of London have an extent of 6588 acres, of 8·8 per cent. of the entire area. They are owned and maintained by the Government, the City Corporation, the London County Council, the various Borough Councils, the Conservators of Putney and Wimbledon Commons, the Metropolitan Public Gardens Association, and various other public bodies and persons. The London County Council and the City Corporation also own and maintain forests, parks, and open spaces outside the county, and in some instances we shall specially refer to them. The Government own and maintain Hyde Park, St James's Park, the Green Park, Kensington Gardens, Regent's Park, Greenwich Park, Woolwich Common, and other smaller spaces. The City Corporation own and maintain Highgate Wood, Queen's Park, Kilburn, within

Highgate Ponds

the county, and Epping Forest, Burnham Beeches, and West Ham Park outside the county. The London County Council are responsible for Battersea Park, Bostall Heath and Woods, Brockwell Park, Clapham Common, Hackney Marsh, Hampstead Heath, Victoria Park, Tooting Common, Wandsworth Common, Streatham Common, Wormwood Scrubs, and many other open tracts. It recently came into possession of Hainault Forest, a most beautiful piece of woodland in Essex; and not a year passes without one or more parks and open spaces either being presented to the public or bought by the London County Council. The various Borough Councils maintain such open spaces as disused burial grounds, recreation grounds, gardens in squares, and small commons.

We can realise what a boon all these parks and open spaces must be to London when we remember that many Londoners can never get far away from their place of work or home all the year round. To thousands of men, women, and children the parks and open spaces in the great city afford their only place of recreation and give them some idea of what the country is like. It has been a great advantage to London to have these open spaces for public resort, for there is no doubt that through them the love of Londoners for flowers and birds has been developed. Although Englishmen have not often been in the front rank as great architects, there is no doubt that they have gained a reputation as landscape gardeners; and in our London parks we may see some good examples of landscape gardening. It has been the aim of those who laid out the parks to make them as natural as possible. A walk in Regent's Park or Kensington Gardens will at once show what beautiful tracts of woodland they are, and what care has been displayed in preserving their natural characteristics. In most of the parks, certain portions have been laid out as flower gardens, and the varied colours of the tastefully-arranged beds form charming pictures. Besides the finest trees and beautiful flowers, the Parks also have the great attraction of bird life, but of that we shall read in another chapter.

In some of the London parks, perhaps, there has been a tendency to make too many straight rows and formal walks, but this cannot be said of the commons, or of such a tract as Hampstead Heath. The commons have a distinct charm in their natural beauty and in their freedom, as opposed to the artificial character and restrictions of some of the parks. These commons are also part of the history of the county, and take us back to the time when the land was tilled in common. Not many years ago, there was a desire to build over these commons; but of late a better spirit is abroad, and now every effort is made for the preservation of open spaces in and around London.

In this western portion of the County of London we find the following are the largest open spaces north of the Thames :—Golder's Hill, Hampstead Heath, Parliament Hill, Ravenscourt Park, Waterlow Park, and Wormwood Scrubs; and south of the Thames:—Battersea Park, Brockwell Park, Clapham Common, Streatham Common, Tooting Common, and Wandsworth Common. These are all under the management of the London

County Council, and we will devote the remainder of this chapter to a brief review of them. The parks owned and maintained by the Government will be considered in the next chapter.

Golder's Hill is a picturesque park of 36 acres, adjoining Hampstead Heath. The grounds have stately trees and some fine specimen shrubs. Near the mansion is a small lake with water-lilies, affording a quiet retreat for moorhens and other waterfowl. A little stream runs through a valley whose sloping banks are covered with grasses and wild flowers. A portion of this valley has been set apart for some red-deer, while another enclosure is reserved for the pea-fowl and an emu.

Hampstead Heath is regarded as the finest of London playgrounds, and at holiday times it is visited by many thousands of people. It has a fine position on the north-western heights of the county, and covers an area of 240 acres. It was acquired in 1871 after much agitation and discussion in the law courts, and was dedicated to the public in the following year. It is very undulating in character, and the portions which are covered with gorse and undergrowth are very picturesque. The most famous view is from Spaniards-road, which crosses the Heath at its highest point. Hampstead Heath is well supplied with water, and the various ponds are used for bathing, fishing, and model-yacht sailing. At various parts enclosures and plantations have been formed as sanctuaries for bird life.

Parliament Hill and Fields adjoin Hampstead Heath, and the surroundings of this fine open space of 267

Hampstead from Parliament Hill

acres are very beautiful. It has been thought by some
writers that the name Parliament Hill suggests that the
place was formerly used for the meeting of the folk-moot.
There is a tumulus known locally as the Tomb of
Boadicea. This, however, was more probably raised by
the Romans as a boundary mark.

Ravenscourt Park is at the western end of Hammersmith and contains an ornamental lake, and an avenue of

Parliament Hill, Hampstead

stately elms. A walled garden has been laid out with old
English flowers and forms a quiet retreat.

Waterlow Park of 26 acres, on the southern slope of
Highgate Hill, was for many years the home of Sir Sydney
Waterlow, who gave it to the London County Council
for a public park. The park is undulating, and has old
cedars and many other well-grown trees and shrubs. Animal

and bird life is encouraged, and the old English garden is
always gay with flowers. This park has interesting historical associations, and Lauderdale House, which has been
restored, dates from the seventeenth century. It takes its
name from the Earl of Lauderdale who lived here, and
there is a tradition that for some time it was the residence
of Nell Gwynn.

Wormwood Scrubs is a great common, 193 acres in
area, on the western border of the county. Part of it
is used for military purposes, and is divided from the
portion to which the public have free access by a belt of
trees.

Battersea Park has an area of nearly 200 acres, and is
the largest municipal park of London. It is on the right
bank of the Thames between the Chelsea and Albert
Bridges, and was formed by the Government in 1846 from
Old Battersea Fields, a low-lying marshy tract. The
chief feature of the park is the sub-tropical garden, which
is planted in the summer months with palms and other
similar plants. Another portion of the park has been
planted with examples of the commoner natural orders
for botanical study, and adjoining this is a garden where
an attempt has been made to naturalise some of the
hardier wild flowers. There is an enclosure for deer,
and a small shelter for owls; while on the lake will be
found many varieties of water-fowl. The river frontage
of the park is about three-quarters of a mile in length,
and affords a promenade with views of Chelsea on the
other side of the river.

Brockwell Park occupies the slope of a hill rising

from the Norwood and Dulwich Roads to Tulse Hill. Its charm is due to its natural beauties, although much has been done to make it useful to the residents in the neighbourhood. The Old Garden was formerly the kitchen garden of the mansion, and is now surrounded by high walls covered with roses and other flowering creepers. The garden is laid out in the formal geometric

The Lake : Battersea Park

style, and the old-fashioned herbs and plants, the quaint sun-dial, and the picturesque well and bucket give the impression of a typical old-world garden. Near the house there is an aviary stocked with pea-fowl, pheasants, doves, and squirrels.

Clapham Common of 220 acres is fairly level, and is much used for games. Streatham Common is situated

at the southern extremity of the county, and from its
higher ridges fine views of the surrounding country are
obtained. The upper part of the common is covered
with gorse, brambles, and other undergrowth, and being
undulating is one of the most picturesque places in the
south of London.

Tooting Common really consists of two commons—
Tooting Bec and Tooting Graveney—which are separated by an avenue of fine trees. It is a large open space
of 217 acres, but suffers from being cut up into three
separate areas by railway lines.

Wandsworth Common has an area of 183 acres, and
forms a small portion of the extensive waste lands that
formerly belonged to the large manor of Battersea and
Wandsworth. Although it is much intersected by roads
and railways, the common has many attractions. A good
deal of planting has recently taken place, and the old
gravel pits have been utilised for the formation of a sheet
of water.

5. The Royal Parks—St James's Park. The Green Park. Hyde Park. Kensington Gardens. Regent's Park.

West London has a larger area of open spaces
and parks than East London, so it is necessary to give
an additional chapter on the Royal Parks which are

owned and maintained by the Government. Westminster has the whole of St James's Park, the Green Park, and Hyde Park; while Kensington Gardens are divided among Kensington, Paddington, and Westminster. Regent's Park is in the boroughs of Hampstead, St Marylebone, and St Pancras.

St James's Park is the most beautiful and aristocratic of the London Parks, for round it are the royal palaces and some of the finest houses. An eminent French writer describes St James's Park as a genuine piece of country, and of English country; with huge old trees, real meadows, and a large pond peopled with ducks and water-fowl; while cows and sheep feed on the grass, which is always fresh. Henry VIII first formed it from a marshy meadow belonging to the Hospital for Lepers. It was replanted and beautified by Charles II, and finally arranged by George IV much as we see it to-day.

On the north side of the park is the Mall, the ancient fashionable promenade of London before Rotten Row became the mode. The Mall has been recently re-constructed in connection with the Queen Victoria Memorial. It is now 200 feet wide, of which space 65 feet in the centre are devoted to the Processional Road from Trafalgar Square to Buckingham Palace. At the eastern end it is entered by a fine triple arch designed by Sir Aston Webb.

St James's Park has charming views of the public buildings at Westminster, and seen through the trees one has glimpses of the grey old Abbey, of

"Cloud-capt towers and gorgeous palaces."

The park is of special interest to the lover of birds, for
the lake is the haunt of a large collection of water-
fowl of many species, whose breeding ground is Duck
Island.

This beautiful park has many historical memories.
Charles I, attended by Bishop Juxon and a regiment of

The Queen Victoria Memorial

foot, walked on January 30, 1648–9 through the park
from St James's Palace to the scaffold at Whitehall. In
this park, Cromwell took Whitelocke aside and sounded
him on the subject of a King Oliver. Some of the trees,
planted and watered by Charles II, were acorns from the
royal oak at Boscobel; and the Merry Monarch kept a
menagerie and some aviaries in Birdcage Walk, the road

which borders the Park on the south. It was a favourite pastime of Charles II to come here with his dogs and feed his ducks.

The Queen Victoria Memorial is at the west end of St James's Park. It is a large semi-circle laid out as an ornamental garden, with architectural and sculptured additions. The central object is the fine monument of the Queen by Sir Thomas Brock, R.A., which is visible from the extreme east end of the Processional Road.

The Green Park is an open area of 53 acres between St James's Park and Piccadilly. Its name well describes the park, for it consists of pleasant greensward, with some shrubberies and flower-beds. In the time of James I much of the area covered by this park was a farm, and it was reduced in size by George III, who annexed part of it to add to the grounds of Buckingham Palace. The road connecting St James's Park with Hyde Park and skirting the garden wall of Buckingham Palace is known as Constitution Hill. Near the upper end of this road, Sir Robert Peel was thrown from his horse and killed; and in this road Queen Victoria was fired at on three occasions.

Hyde Park is reached from the Green Park by crossing Piccadilly. It is one of London's great lungs, and has an area of 364 acres. The park is entered from Piccadilly by a triple archway designed by Decimus Burton, and erected in 1828. The name is derived from the Hyde, an ancient manor of that name, which belonged to the abbots and monks of Westminster till the dissolution of the religious houses by Henry VIII. It then

became the property of the Crown, and for much of its
present beauty it is indebted to William III and Caroline,
wife of George II. It was Queen Caroline who formed
the sheet of water called the Serpentine, and the carriage
drive along the north bank is called the "Lady's Mile."
The bridle road running from Apsley House to Kensington Gardens is Rotten Row, probably a corruption

Hyde Park, the Serpentine

of *Route du Roi*—King's Drive. The flower-beds are
always an attraction, and the rhododendron show in June
is specially famous. The entrance to the park from
Oxford Street was by the Marble Arch, which was moved
from Buckingham Palace in 1851, and re-erected here.
It is a triumphal arch in the style of the Roman Arch
of Constantine, and its bronze gates are admirable.

Recently it has been found necessary to make an open space around the Marble Arch, and this has been done by setting back the Park entrances.   Now the Marble Arch is quite isolated and meaningless, for it is no longer the entrance to the Park.

In 1851, the Crystal Palace, or Great Exhibition Building, covered nearly 19 acres on the south side of

The Broad Walk, Kensington Gardens

Hyde Park, and near its site rises the Albert Memorial, the national monument to the Prince Consort.   It is a gothic cross or canopy designed by Sir Gilbert Scott, and its spire reaches a height of 175 feet.   An enormous amount of money—£120,000—was lavished on it, but its success as a work of art is much questioned.

Kensington Gardens are continuous with Hyde Park,

but there is a great difference between them, for the Gardens are more rural and have much finer trees. A recent writer on London well says that Kensington Gardens are a paradise of lovely sylvan glades and avenues, hardly less picturesque than St James's Park. Many other writers have paid their tribute to these gardens. Matthew Arnold felt their charm, and in one of his sonnets he writes :—

> "In this lone, open glade I lie,
>   Screen'd by deep boughs on either hand;
>   And at its end, to stay the eye,
>   Those black-crowned, red-boled pine trees stand."

Kensington Gardens were laid out in the reign of William III, and originally consisted of only 26 acres. Other monarchs have added to them, and now they have an area of 275 acres. The bridge over the Serpentine separating the Gardens from the Park was designed by Rennie, and erected in 1826. Adjoining Kensington Gardens on the west is Kensington Palace, of interest as a royal residence, and specially noteworthy as the birthplace of Queen Victoria.

Regent's Park is the largest of the London parks, being 472 acres in extent. It is part of old Marylebone Farm and Fields, and was laid out in 1812 from the plans of Mr John Nash. The Park derives its name from the Prince Regent, afterwards George IV, who intended building a residence here, and Regent Street was designed as a communication from it to Carlton House. Pleasant paths run in every direction over Regent's Park, and the two principal roads, the only

carriage drives, are called the Outer and the Inner
Circle. The former, two miles in length, encircles the
park, while the latter, in the middle of it, encloses the
Botanical Gardens. These are very pretty, and in May
and June large flower-shows are held in them. They
have attached to them an interesting museum and

Regent's Park

collections of orchids and sea-weeds. The chief road
in the Park for pedestrians is the charming Broad Walk,
which extends from Park Square on the south to
Primrose Hill on the north. It is bordered with trees,
and the southern portion is laid out with beds of flowers.
The lake in the western half of the park is picturesque
and has numerous water-fowl. The chief attraction of

Regent's Park is in the Zoological Gardens, the most complete in the world. The Zoological Society was founded in 1826, mainly by Sir Humphry Davy and Sir Stamford Raffles, and the present Gardens were opened in 1828. Darwin and Huxley studied here, and the Gardens now contain a very full series of vertebrate animals. On the east side of the park is St Katherine's Hospital, with its chapel. Founded centuries ago, it stood originally near the Tower, but was removed here in 1827, to make room for St Katherine's Docks. Separated from Regent's Park by two roads and the canal rises Primrose Hill, which has been planted and laid out with walks. From its summit may be gained extensive views over London.

6. The River Thames. The Embank- ment. The Wandle. The Bridges.

In an early chapter we read that London was founded on a site about 60 miles from the coast, and we also learnt that "London" and "Thames" are the only Celtic words remaining in this area to remind us of the British occupation of our country. Now, as the Thames has played such an important part in the growth and development of London, it will be necessary to devote a little time to the study of this, our greatest river. A recent writer has said that "The river has made London, and London has acknowledged its obligations to the Thames. It was the Silent Highway along which the chief traffic of the City passed during the Middle Ages....The river continued to be the Silent Highway until the nineteenth century, when it lost its high position. With the construction of the Thames Embankment the river again took its proper place as the centre of London, but it did not again become its main artery."

The Thames, indeed, with its tides and its broad shining waters, has always been the source of London's wealth, and has been well named by one poet " Father Thames," and by another writer the "Parent of London." Throughout our history and literature the Thames plays a prominent part, and we shall find in the pages of this volume many references to it. With English poets it has been a favourite theme, and we find such expressions as " The silver-streaming Thames " of frequent occurrence, while Denham has sung its praises in some noble couplets :—

"O could I flow like thee, and make thy stream
My great example as it is my theme:
Though deep yet clear, though gentle yet not dull,
Strong without rage, without o'erflowing full."

The watermen of London were long famous, and many were the sports on the Thames that gave colour to the life of Londoners. There are many records of the Thames being frozen over in severe winters, and some of the Frost Fairs on the ice were of considerable duration. One of London's historians of the sixteenth century gives us some idea of the plentifulness of the fish caught in the Thames by London. "What should I speak," says Harrison in 1586, "of the fat and sweet salmons, daily

Fair on the Thames, February 1814

taken in this stream, and that in such plentie as no river in Europe is able to exceed it?" The first salmon of the season was generally carried to the King's table by the fishermen of the Thames. A sturgeon caught below London Bridge was carried to the table of the Lord Mayor; if above bridge, to the table of the King or Lord High Admiral.

London has had great pageants on its river, and in Stuart times the Dutch ships were brought on its stream almost within gunshot of the Tower. Queen Elizabeth died at Richmond, and her body was brought in great pomp by water to Whitehall. Nelson's body, too, was carried in great state by water from Greenwich to White-hall. Many a state prisoner, committed from the Council Chamber to the Tower, was taken by water: and we all remember that striking scene in our history, when the Seven Bishops were carried by the Thames to the Tower. Almost as a sequel to the last event, James II himself fled from London by water, and, in his flight, threw the Great Seal of England into the Thames.

Such, then, are a few of the historic landmarks which draw our attention to the river that has made London the capital of the British Empire.

It will be seen by a reference to the map that the Thames divides London into two unequal portions. It is navigable and tidal throughout its course through London; and from its source in the Cotswold Hills to the Nore the direct length is 120 miles, although with the windings it is probably 220 miles in length.

The Thames, or Tamesis as it was once called, is the

The Thames from Richmond Hill

earliest British river mentioned in Roman history. Its name, as we have seen, is of Celtic origin, and its derivation is probably the same as that of the Tame, the Teme, and the Tamar in other parts of England. The upper part of the main stream is often called the Isis, and not the Thames, until it has received the waters of the Thame near Dorchester in Oxfordshire. In its upper course it passes through some of our finest agricultural country, while below London Bridge it is one of the most important commercial highways in the world. The Thames begins to feel the tide at Teddington, and from there to the Nore, a distance of 68½ miles, the tide ebbs and flows twice in the day. The force of the tide is very great, and its power can be seen at Blackfriars Bridge, where the water swirls round the piers and rushes through the arches like a mill-race.

The Thames enters the County of London at Hammersmith, where a bridge crosses the river to Barnes on the right bank. From this bridge westwards, almost to Chiswick, the riverside is known as the Lower Mall, and the Upper Mall. Some of the old houses along this portion of the Thames are now the headquarters of boating and sailing clubs. The Lower Mall is separated from the Upper Mall by a little creek spanned by a wooden foot-bridge. Near this is "The Doves," a little old-fashioned inn, but formerly a coffee-house, where the poet Thomson is said to have written his *Winter*. On the western side of the Lower Mall is Kelmscott House, intimately associated with William Morris, poet, craftsman, and socialist. Morris named his house on the

Upper Mall after his residence on the Upper Thames. Mr Mackail, in his *Life of Morris*, says that "the

"The Doves"

hundred and thirty miles of stream between the two houses were a real as well as an imaginative link between them. He liked to think that the water which ran under his windows at Hammersmith had passed the meadows and gray gables of Kelmscott; and more than once a party of summer voyagers went from one house to the other, embarking at their own door in London, and disembarking in their own meadow at Kelmscott."

From Hammersmith Bridge the Thames bends and makes a semi-circular curve in which Fulham is enclosed. On the right bank are the West Middlesex Reservoirs and the grounds of the Ranelagh Club. On the left bank are Fulham Palace and Bishop's Park, and a little to the east the Thames is spanned by Putney Bridge. Putney is the headquarters of many rowing clubs and presents a scene of great activity in the season. It is also the starting-point of the race which is rowed every year between Oxford and Cambridge. The Boat Race as it is always called, without further definition, has been rowed since 1839, and is one of our national institutions.

A little distance eastward of Putney Bridge, by the riverside, is Hurlingham House, with spacious grounds, beautiful gardens, and a lake of four acres. Here it is that the Hurlingham Club has its polo, tennis, and other sports, which attract large numbers of the upper classes in the season. Before the river turns north it is crossed by Wandsworth Bridge, and onwards to Battersea Bridge this section of the river is known as Battersea Reach. On the left bank there is Chelsea with its Embankment and its avenue of plane trees, and the Royal Hospital, one of the most stately of Wren's buildings. Across the river is the greenery of Battersea Park,

Hammersmith Bridge

(The Oxford and Cambridge Boat Race)

which is the finest of the south London open spaces. Before Vauxhall Bridge is reached we see a Floating Fire Brigade Station close to the left bank. The river now runs to the north as far as Charing Cross Bridge, and on the right bank the Albert Embankment extends from Lambeth Bridge to Westminster Bridge, and has converted the river front into a fine promenade. On the Lambeth side there is Lambeth Palace with its square castellated towers of brick, which have toned to a lovely dull deep red; and St Thomas's Hospital with its seven great blocks connected by arcades. On the Westminster side there are the National Gallery of British Art (better known as the Tate Gallery), the Abbey, and the Houses of Parliament.

At the head of the long flight of steps leading down to the Thames at the Westminster end of the Victoria Embankment is Thornycroft's "Boadicea," a spirited statue of the British Queen and her daughters in a scythed chariot. The Thames Embankment extends from Westminster Bridge to Blackfriars Bridge, and is one of the greatest London improvements in the nine-teenth century. This work, begun in 1864 and not completed till 1870, was planned by Sir Joseph Bazalgette. Now the Victoria Embankment is one of London's finest thoroughfares, and on the landward side has stately public buildings, such as Scotland Yard and Somerset House, fine hotels, well-planned gardens with statues of great men, and the historic Temple buildings. Between Charing Cross Railway Bridge and Waterloo Bridge, which have been styled the ugliest and the

The Statue of Boadicea on the Thames Embankmen

handsomest of the London bridges, there stretches the
finest of the Embankment Gardens, and in their north-

The Water Gate, Embankment Gardens

west corner is the Water Gate of York House, a London
relic of singular interest and charm.   It is variously

attributed to Inigo Jones and Nicholas Stone, and its presence helps us to realise what a difference has been made to the river by the Embankment, for it is now several hundred feet from the river, whose waters once reached the steps at its base. Here it was that the Thames watermen, in top hats and full-skirted red coats, used to land passengers from their great barges. Between the river and the gardens is Cleopatra's Needle, a red granite monolith with its mystic emblems carved by Egyptian sculptors more than 3000 years ago.

The right bank of the Thames from Westminster to Blackfriars Bridge has little of interest. The lofty Shot Tower rises amid wharves and riverside buildings which are in marked contrast to the fine buildings along the left bank. What strikes one most is the life on the river, which seems ever on the move, with its numerous coal barges and every kind of river craft. The eastern section of the Thames, from the City boundary to the point at Woolwich where it ceases to be a river of the County of London, belongs to the volume on East London, which also deals with the Port of London.

There is one little tributary of the Thames on the right bank which may be noticed here, before we consider the bridges. The Wandle, which rises near Croydon in Surrey, flows eastward by Beddington and Carshalton, and then northward past Morden, Merton, and Tooting to Wandsworth, to which place it gives its name. Here it enters the Thames after a course of about ten miles. It is worth noting that calico-bleaching and printing were formerly carried on along this little river Wandle, and that at Wandsworth there were numerous factories for this purpose.

The Thames as it flows through the County of London is crossed by many fine bridges, from Hammersmith Bridge in the west to the Tower Bridge in the east. And yet less than two centuries ago London Bridge was the only bridge over the Thames in London. It is still the most important, for it connects the City, the centre of London's business, with Southwark on the Surrey side of the Thames. London Bridge, Blackfriars Bridge, Southwark Bridge, and the Tower Bridge are maintained by the City Corporation and fall within the volume that deals with East London. In this chapter we will consider the bridges to the west of Blackfriars Bridge, all of which are maintained by the London County Council.

Beginning at the extreme west, Hammersmith Bridge is the first that claims our attention. It is an iron suspension bridge, with a span of 400 feet, and connects Hammersmith with Barnes. It was completed in 1885, and superseded the original bridge of 1827.

Putney Bridge is a handsome structure of granite which was opened in 1886. It displaced an old wooden structure of 1729 which had superseded the ferry. The present bridge was designed by Sir J. Bazalgette who was the engineer for the Thames Embankment.

Wandsworth Bridge merely requires to be mentioned. Chelsea Bridge was built in 1858 on the suspension principle by Thomas Page, and is considered the most graceful of recent Thames bridges. The Albert Bridge was opened in 1873, and Battersea Bridge in 1890 took the place of a picturesque old wooden bridge which had superseded the ancient ferry at this spot.

Vauxhall Bridge, the work of Mr Fitzmaurice, is a structure of iron and steel. It was opened in 1906, as the successor of an old bridge which had done duty since 1816. Lambeth Bridge is narrow and mean-looking, and was built for a company in 1862–3. References are often found in old writers to Lambeth Bridge, but that was simply a landing place, for, until the present bridge was opened, the only way of crossing the river at this point to Westminster was by ferry—the old Horseferry.

Westminster Bridge is one of the finest bridges in London. It has seven arches of iron resting on granite piers, which are 30 feet below low water. The parapets and ornamental portions were designed to harmonise with the Houses of Parliament, and the roadway is said to be wider than that over any other bridge in the world. This handsome structure, built from designs made by Mr T. Page, was opened in 1863. The bridge commands a fine view of the Houses of Parliament, and of St Thomas's Hospital. The earlier bridge was the first bridge over the Thames at Westminster, and was opened in 1750.

Charing Cross Bridge carries the South Eastern and Chatham Railway across the Thames. It was designed by Sir John Hawkshaw and took the place of Hungerford Suspension Bridge, which, after standing here for some years, was removed in 1863 to Clifton near Bristol, where it forms a striking object. The bridge at Charing

Waterloo Bridge

Cross is a disfigurement to the Embankment, and its massive iron pillars have caused it to be called the ugliest bridge in London. One of Whistler's most noteworthy etchings is of this bridge.

Waterloo Bridge, the work of the elder Rennie, was considered by Canova the noblest in the world. It was begun in 1811, and opened in 1817, on the second anniversary of the battle of Waterloo from which it takes its name. It is of granite, with nine semi-elliptical arches, of which the most northerly stretches across the Embankment.

7. Rivers of the Past. The West=bourne and the Tybourne, or Tyburn.

We may now proceed to consider the rivers and streams that once fell into the Thames. The most important of those on the north were the Westbourne, the Tybourne or Tyburn, the Holebourne or Fleet or Wells River, and the Walbrook. Not one of these streams now runs above ground, and if they flow, it is merely as underground sewers. Their courses, however, can still be traced by the names of places that formerly stood on their banks. As regards the streams of the south, they have little bearing upon the history of ancient London. The chief fact to bear in mind about the south was the vast extent of marsh-land, now covered with thousands of houses. A few streams crossed the marshes,

among them the Falcon Brook and the Effra. In this
volume, we will consider the Westbourne, and the
Tybourne or Tyburn.

The Westbourne rose in the Hampstead Heights,
and after crossing the site of the present Edgware Road

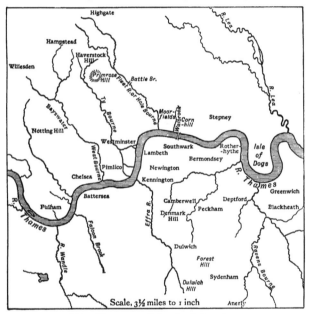

The Streams of Ancient London

spread out into a shallow water, which probably gave the
name to Bayswater. When Hyde Park was laid out in
1733, this stream was used to form the Serpentine. It now
flows underground, and leaving the Park at Albert Gate,
falls into the Thames through the Ranelagh Sewer.

The Tybourne, or Tyburn, rose in Conduit Fields
on the slope of Hampstead. Thence it ran for a few
hundred yards through Regent's Park, and the present
Marylebone Lane marks its windings for us. The stream
flowed onwards to Piccadilly, and across the Green Park.
Mr Loftie says that "the windings of the Tyburn are
occasionally revealed by a line of mist, which shows that
it has not been wholly dried up in its underground
course." Near Buckingham Palace it divides into three
branches. Part falls, or used to fall, into the Thames
through the ornamental water in St James's Park ;
another part ran into the ancient Abbey buildings for
the use of the monks; and the third branch passes under
the Palace grounds and falls into the Thames at the
King's Scholars' Pond Sewer. The Tyburn has entered
into the historical associations of London. Tyburn was
the original name of the present St Marylebone. Tyburnia
is the district that borders on the Bayswater Road.
Tyburn Lane was the ancient name of Park Lane, and
Tyburn Road was the old name for Oxford Street. The
Tyburn also gave its name to the place of execution for
criminals convicted in the county of Middlesex. Tyburn
Gallows, or Tyburn Tree, existed as early as the reign of
Henry IV, and stood at the south end of the present
Edgware Road. The Gallows was a triangle in plan,
having three legs to stand on. In Tarlton's *Jests*, 1611,
there is this remark, "It was made like the shape of
Tyborne, three square." Many celebrated and notorious
characters were executed at Tyburn, and such names as
John Felton the assassin of Buckingham, Jack Sheppard,

Jonathan Wild, and Dr Dodd at once rise in our memory. The last person executed at Tyburn was John Austin, on November 7, 1783; and the name of Ketch, one of the executioners in the seventeenth century, is now synonymous with hangman. On the three wooden stilts of Tyburn the bodies of Oliver Cromwell, Ireton, and Bradshaw were hung, on the eleventh anniversary (January 30, 1660–1) of the execution of Charles I. Their bodies were dragged from their graves in Westminster Abbey, and removed at night to the Red Lion Inn in Holborn, from whence they were carried next morning in sledges to Tyburn.

8. The Water-Supply of London—Past and Present.

The Water-Supply of a great city is of the utmost importance, for on its good quality and constant flow depend largely the health and happiness of its people. In considering the supply of water to London, it is well to remember that ancient London and the many parishes now comprised in the modern county, arose on sites where a supply of good drinking-water could readily be obtained from natural springs and brooks, or by means of wells. The earlier settlements were made on the tracts of gravel and sand, and thus the growth of London was regulated for a long period by the distribution of these water-bearing strata. Thus we find that the City expanded westwards to Chelsea, Kensington, and Hammersmith ;

southwards to Clapham, and Camberwell; eastwards
to Bow, and Hackney; and northwards to Islington.

View of the Conduit at Bayswater

Such districts as Camden Town, Kentish Town, and
Kilburn were not populated until a supply of drinking-
water from a distance was brought in conduits.

It is a matter of history that, from 1680–1840, some of the London wells and springs attained fame as wells and spas. Thus we read of Beulah Spa, Bermondsey Spa, Islington Spa, Holywell, Clerkenwell, and Sadler's Wells. At Well Walk, Hampstead, a chalybeate spring was utilised until quite recently. The first conduit for the supply of water to London was that of Tyburn, which was completed in 1239, when water was conveyed in leaden pipes to the City. Much water was also obtained in buckets from the river, and in 1582 the supply was facilitated by means of water-wheels attached to the arches of old London Bridge. After a time wooden conduits were used, and a more extended system of supply to houses was introduced. In opening some of the London streets it is no uncommon thing to find these wooden conduits as they were placed a long time ago. Some of these relics of the early times of London's water-supply may be seen in various museums.

With the growth of London, the supplies of water from the gravel soils became contaminated, and the water of the Thames near London Bridge was very bad. From the close of the seventeenth century and onwards to 1855, companies were formed for taking water from the Thames near Charing Cross, and higher up; but since that year, no water has been drawn by any company from the Thames below Teddington Lock.

Sir Hugh Myddelton was the pioneer in bringing water to London from a distance. In 1608, he commenced the cutting of the New River, and five years later that artificial channel was completed. As a result

of his efforts the New River Company was formed in
1619, and down to the present century it has been of the
greatest service to London, in supplying an abundant
quantity of excellent water from the river Lea and
from springs in the Chalk, as well as from deep wells
sunk into the Chalk.

Many artesian wells have been sunk through the
London Clay into the Lower Tertiaries and Chalk, and
since 1790, breweries and other large establishments have
used this means of obtaining their water. One of the
deepest borings through the Chalk in the London Basin
is that at Kentish Town, where various beds have been
passed through to a depth of over 1300 feet.

The water-supply of London and the surrounding
districts is now controlled by the Metropolitan Water
Board. Before the year 1902, this great area was supplied
by eight London Water Companies; but by an Act passed
in that year the Water Board was created for the purpose
of purchasing and managing the undertakings of those
companies. The Metropolitan Water Board area is much
greater than that of the County of London and extends
from Ware in Hertfordshire to Esher in Surrey, and from
Romford in Essex to Chevening in Kent.

The Metropolitan Water Board in 1908 had to supply
a population of more than 7,000,000 persons, and to
deliver a daily average of 224,000,000 gallons. The
whole of this water is obtained from the Thames, the
Lea, and from various springs and wells in the locality.
The water mains have a total length of 6041 miles, and
the water supplied to the great population of London and

its environs is of a very high standard of excellence and of purity.

The great fault of the water supplied over this area is " hardness," that is, the containing a quantity of bicarbonate of lime, yet it is well known that many of the healthiest districts are those with hard water. It is this hardness of chalk waters which furs our kettles, and wastes our soap; but it is not easy in the London area to obtain other than hard water, for much of it is derived from wells in the Chalk.

9. Geology.

In geological language, London is said to be situated in a " basin "—the " London Basin." This basin has been carved out of strata belonging to the early Tertiary period, which is called Eocene. The solid foundation is composed of the Chalk, a formation here about 600 feet in thickness. This it is which really constitutes the London Basin, whose broad rim comes to the surface in the Chiltern Hills in the north and north-west, and in the North Downs in the south. Although the Chalk may be called the basement-rock of the London Basin, yet wells have been sunk in the middle of that area to so great a depth as to pass through it into the beds below.

For the purpose of this chapter we may begin our description of the rocks under London with the Gault, the formation that occurs almost universally in the London Basin. The Gault is a marine deposit, and seems to have been deposited in a moderately deep, quiet sea, and not along a shore-line. It is a bluish clay and varies in thickness from 130 to 200 feet at such deep borings as those at Tottenham Court Road, Kentish Town, and Mile End.

Above the Gault is the Upper Greensand, which is also of marine origin. It consists of clayey greensand varying in thickness from 10 to 30 feet. In all cases, it has been found in the same deep borings as the Gault.

The Chalk comes after the Gault, and is perhaps the best known rock in England. In the deep borings

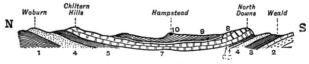

The London Basin

1.	Oxford Clay.	6.	Upper Greensand.
2.	Hastings Beds.	7.	Chalk.
3.	Weald Clay.	8.	Lower London Tertiaries.
4.	Lower Greensand.	9.	London Clay.
5.	Gault.	10.	Bagshot Beds.

through the Chalk in London, the thickness of this formation varies from 645 to 671 feet. By its fossils the Chalk is proved to be the deposit of a fairly deep sea, something of the same character as that now forming in the mid-Atlantic Ocean. The Chalk is not much exposed in the County of London, but there are some pits near Deptford and also near Lewisham where it may be seen.

Over the Chalk come a number of thin but varying

beds known as the Lower London Tertiaries. To each of these divisions a local name has been given. The Thanet Beds are the lowest, and are so named from the fact that they are the only Tertiary formation in the Isle of Thanet. The Woolwich and Reading Beds succeed, and are named from their occurrence in the neighbourhood of those places. The upper formation of Oldhaven Beds follows the last, receiving its name from the good section of it which may be seen at Oldhaven Gap, near Reculver in Kent.

The Thanet Beds consist almost wholly of fine soft sand, very pale grey or buff, and very compact. They are without pebbles and without fossils, and have a thickness of about 30 feet. Sections rarely reach the surface in the County of London, but the Thanet Sand may be well seen at Plumstead, Woolwich, and Lewisham.

The Woolwich and Reading Beds are a group of clays and sands having a thickness of 60 feet or less. Shells of an estuarine character are found in the clay, and this fact proves that the beds were deposited at or near the mouths of streams. Sections of these beds may be seen at Woolwich, Charlton, and Lewisham.

The Oldhaven and Blackheath Beds consist mostly of a bed of perfectly-rolled flint pebbles, in a base of fine, sharp, light-coloured sand. The thickness is as much as 50 feet, and fossils are met with in parts. Interesting open sections of the Blackheath Beds may be seen at Eltham, Bostall Heath, Woolwich, Plumstead Common, and Blackheath.

Overlying the last beds is a great mass of clay known

as London Clay. Although this formation takes its name
from the Metropolis, it is well to remember that it extends
from Marlborough on the west to Yarmouth on the
north-east. In the neighbourhood of London the Clay
is 400 feet thick, and many fossils have been found in it.
The London Clay forms a very broad band right through
the London area from south-west to north-east, and
excellent sections may be seen at Plumstead Common,
Hampstead, and Highgate.

Above the London Clay come a group of sands
which may be comprised under the name Bagshot Beds.
As a whole they form a more or less barren sandy tract
of rising ground, which is partly open, but sometimes
covered with fir and larch. The hills of Hampstead and
Highgate are perhaps the most prominent heights in north
London, and although the Bagshot Beds cap these hills,
it must be remembered that their longer slopes are of
London Clay.

In the London Basin, after the Bagshot Beds, we
come to a great gap in the series of geological formations.
The beds just named are Eocene, and we find nothing more
till we are almost out of the Pliocene Period. Gravels,
sands, and clays are found at various levels down nearly
to the present level of the Thames, and this newer set
of deposits may be classed under the term Drift. The
Boulder Clay is one of the most important of these
deposits. It is stiff and tenacious, and often studded with
pieces of Chalk. Good sections of the Boulder Clay are
rare but it can sometimes be seen in temporary openings
and in roadside sections and ponds.

Passing over some deposits known as Brick earth, and Valley or River Gravel, we come to the last and newest deposits of the district. These Alluvial Deposits are confined to the very bottoms of the valleys in which rivers run. They comprise the strip of marsh-land or Alluvium, which fringes the river over small areas above London, and over broader tracts in southern Essex and northern Kent. The Alluvial Deposits are from 12 to 20 feet thick, and the old river mud often contains bones of the ox, deer and elk, as well as implements of stone, bronze, and iron.

It is advisable, while reading this chapter, that constant reference should be made to the geological map at the end of this volume. The reader is also advised to pay a visit to the Geological Museum at Jermyn Street, where there is a large model of London and the neighbourhood. It is on the scale of 6 inches to the mile, and represents an area of 165 square miles, and owing to its great size it is in nine sections. It gives an excellent idea of the geological structure of London.

10. Natural History.

Among all the English counties, London has least to attract the lover of natural history. Almost its entire area is given over to bricks and mortar, and outside the parks and open spaces there are few places where the flora and fauna can be studied in the same way as in the neighbouring counties of Essex, Kent, and Surrey. There are tracts in the north and in the south-east and south-west where the population is not so dense as in central London, but even in those districts streets are being made every year, with the consequent destruction of plant and animal life.

It would almost be easier to write about the trees and flowers in London many years ago than of those at the present time. London was once famous for its trees and flowers. Vinegar Yard, Covent Garden, was the vineyard of Covent Garden. Saffron Hill was once covered with saffron (*Crocus sativus*). The red and white roses of York and Lancaster were plucked in the Temple Gardens; and Daniel, the poet, in the reign of Elizabeth, had an excellent garden in Old Street, St Luke's. Gerard the herbalist in the same reign had a choice assemblage of botanical specimens in his garden at Holborn.

The flora of the south-east and south-west of the county of London is similar to that of Surrey and Kent. The blue wood anemone (*A. apennina*) was formerly abundant as an introduced plant in Wimbledon Park, but is now extinct. The cowbane (*Cicuta virosa*) formerly grew by the Thames at Battersea. Gerard (1633) records that this plant grew in Moor Park, Chelsea, but of course it is no longer found there. The sea-aster (*A. Tripolium*) grew by the Thames near Battersea, but is no longer a plant of the county.

Furze, broom, briars, bracken, and heath are abundant on Barnes Common, and on Putney Heath and Wimbledon Common are to be seen scrub of stunted oak, hazel, birch, and sallows, with plenty of tall furze. Wandsworth Common now grows nothing unusual, but formerly its speciality was the water-soldier (*Stratiotes aloides*), while Streatham Common was famous for *Senecio viscosus*.

Sir Hans Sloane, M.D.

The Plumstead marshes and the flats below Woolwich and towards Erith have now been drained and put under pasturage. Aquatic plants, both rare and ordinary, grew here in great variety and abundance, but they have now almost disappeared. In the north of the county

5—2

Hampstead Heath is the chief open space.   The ground is there broken into pits and hillocks, and much bracken grows, with a few white and black thorns.

There is one survival in London which is worth a passing notice.   Adjoining the Chelsea Embankment is the Physic Garden, founded in 1673 and presented by Sir Hans Sloane in 1722 to the Society of Apothecaries on condition that 50 new varieties of plants grown in it should be annually furnished to the Royal Society, until the number so presented amounted to 2000.   It was famed for its fine cedars, the first to be grown in England.   They are now no longer existent, the last being felled in 1904.

The great Linnaeus visited this garden in 1736, and Kalm the Swedish naturalist in 1748.   Towards the end of the last century, the Apothecaries' Society being no longer desirous of maintaining the Garden, it was vested in the London Parochial Charities in 1899.   A committee of management was appointed, and as a result of their work many improvements have been made.   The green-houses have been rebuilt, and a well-fitted laboratory and lecture room erected.   The place is now used by students of the Royal College of Science, and members of various schools and polytechnics.   Courses of advanced lectures in Botany are arranged by the University of London; and specimens are supplied to the principal teaching and examining bodies in the Metropolis.

It is very noticeable when houses are cleared away in London, and the space remains derelict for a time, that all kinds of plants will grow and flourish.   In 1909 there was a large tract of waste ground on the north of the Strand, and in the month of July the area was a blaze of purple. Amid the mass of vegetation that had sprung up there were elder, cherry, loosestrife, willow herb, a great crop of rape, and here and there a thistle. In a short space of time, it was not difficult to find as many as twenty different plants, and the question arose how they came to grow in such an unlikely spot. Air-borne seeds such as dandelion settled in this area and grew with little difficulty. Other seeds were carried to Aldwych in the hay or chaff given to horses at the time when the buildings were being demolished and the débris carted away, while the elder and cherry were probably brought by birds.

The parks and open spaces of London are the haunt each year of a far larger number of wild birds than is generally supposed. It is only during the great migration times in spring and early autumn that a full idea can be obtained of the large variety of species which pause for a few hours in the chief London parks as they seek or leave their summer nesting-places. Among these passing visitors to London are the wheatear, the redstart, the sandpiper, the kingfisher, and the great crested grebe. The wheatear and the brilliant kingfisher are by no means unknown in Hyde Park and Kensington Gardens.

Besides these birds, which in London are essentially birds of passage, there is a large group of birds which are regular residents for the whole or part of the year, and thus save London from the reproach of being a birdless county. There are some birds which find greater security

in London than in the country. One of the most conspicuous examples of this class is the carrion crow, which
is common in some of the parks and squares, more
especially in Kensington Gardens. The crow in London
maintains a steady hostility to rooks, and the Gray's Inn

Sea Gulls on the Embankment

rookery has been harried of late years by the crows, so
that this famous colony is gradually diminishing and
disappearing. In 1836, there were 100 nests of rooks
in Kensington Gardens, but owing to the terrorism of
the crows, the rookery there is well-nigh extinct. The

brown owl is another bird of prey which seems glad to find refuge in London. It is not uncommon among the old trees in the larger London gardens, planted a century or more ago, and it has adapted itself to the London life of to-day.

The sparrow is looked upon as a London bird and has every equipment for the needs of London life. He is there viewed with a tolerance, and even with a sentimental affection, which is not extended to him in the country. In London, he is ubiquitous, and seems to find a satisfaction in placing his nest in the most ridiculous positions. The song-thrush and blackbird often sing more vigorously in London during the winter than in most places in the open country. Swallows, martins, and sand-martins are sometimes seen in considerable numbers on the Round Pond, and on the Serpentine, especially in cold and frosty Aprils. The spotted fly-catcher attempts to nest every year in Hyde Park and Kensington Gardens, and the reed-warbler occurs on the London list, for it sometimes pauses on migration in the thickets of reeds at the head of the Serpentine.

Perhaps the most striking example of a recent addition to the birds of London is the annual winter visit of the black-headed gulls which haunt the Thames and the ponds in the parks from October until March. The flocks include from time to time a few gulls of other kinds, the herring gulls being commonest. The gulls first visited London in large numbers in the hard frost of 1895, and have never since abandoned it. Still more remarkable, however, is the introduction of that most wary of all country birds, the wood-pigeon. Unknown a few years ago not only in London but in its near neighbourhood, they became established about 1900 and have remarkably increased. They have become extremely tame and may be seen feeding in dozens in the parks, and even in the roads.

The ornamental waters in the parks are so well stocked with different breeds of duck that it is impossible to say to what extent they are frequented by genuine wild-fowl. There is no doubt, however, that such visitants are numerous. There are upwards of 400 wild-fowl in the splendid collection at St James's Park, which have their breeding place on Duck Island. The herons, whose wings are clipped, have been there three years, and besides black and white swans there are many sorts of geese, wigeon, teal, and mallard.

There was a time when the kite was as familiar in London's streets as the sparrow is now. The swifts used to circle and glide over what is now the densest part of the City, and not a hundred years ago woodcocks were shot in Piccadilly.

11. Climate and Rainfall. Greenwich Observatory and its Work.

The climate of a country or district is, briefly, the average weather of that country or district, and it depends upon various factors, all mutually interacting, upon the latitude, the temperature, the direction and strength of the winds, the rainfall, the character of the soil, and the proximity of the district to the sea.

The differences in the climates of the world depend mainly upon latitude, but a scarcely less important factor is this proximity to the sea. Along any great climatic zone there will be found variations in proportion to this proximity, the extremes being "continental" climates in the centres of continents far from the oceans, and "insular" climates in small tracts surrounded by sea. Continental climates show great differences in seasonal temperatures, the winters tending to be unusually cold and the summers unusually warm, while the climate of insular tracts is characterised by equableness and also by greater dampness. Great Britain possesses, by reason of its position, a temperate insular climate, but its average annual temperature is much higher than could be expected from its latitude. The prevalent south-westerly winds cause a movement of the surface-waters of the Atlantic towards our shores, and this warm-water current, which we know as the Gulf Stream, is the chief cause of the mildness of our winters.

Most of our weather comes to us from the Atlantic. It would be impossible here within the limits of a short chapter to discuss fully the causes which affect or control weather changes. It must suffice to say that the conditions are in the main either cyclonic or anticyclonic, which terms may be best explained, perhaps, by comparing the air currents to a stream of water. In a stream a chain of eddies may often be seen fringing the more steadily-moving central water. Regarding the general north-easterly moving air from the Atlantic as such a stream, a chain of eddies may be developed in a belt parallel with its general direction. This belt of eddies or cyclones, as they are termed, tends to shift its position, sometimes passing over our islands, sometimes to the north or south of them, and it is to this shifting that most of our weather changes are due. Cyclonic conditions are associated with a greater or less amount of atmospheric disturbance; anticyclonic with calms.

The prevalent Atlantic winds largely affect our island in another way, namely in its rainfall. The air, heavily laden with moisture from its passage over the ocean, meets with elevated land-tracts directly it reaches our shores—the moorland of Devon and Cornwall, the Welsh mountains, or the fells of Cumberland and Westmorland —and blowing up the rising land-surface, parts with this moisture as rain. To how great an extent this occurs is best seen by reference to the accompanying map of the annual rainfall of England, where it will at once be noticed that the heaviest fall is in the west, and that it decreases with remarkable regularity until the least fall is reached on our eastern shores. Thus in 1907, the maximum rainfall for the year occurred at Llyn Llydaw copper mine in Carnarvonshire, where 196·16 inches of rain fell; and the lowest was at Clacton-on-Sea, with a record of 16·66 inches. These western highlands, therefore, may not inaptly be compared to an umbrella, sheltering the country further eastward from the rain.

The above causes, then, are those mainly concerned in influencing the weather, but there are other and more

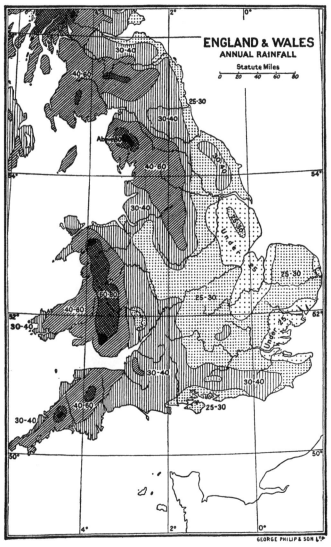

ENGLAND & WALES
ANNUAL RAINFALL

Statute Miles
0   20   40   60   80

30-40
40-60
25-30
30-40
Abpreo
40-60
30-40
60-80
40-60
30-40
25-30
Under 25
Under 25
25-30
30-40
30-40
30-40
Under 25
40-50
30-40
25-30

GEORGE PHILIP & SON LTᴰ

(The figures give the approximate annual rainfall in inches.)

local factors which often affect greatly the climate of a place, such, for example, as configuration, position, and soil. The shelter of a range of hills, a southern aspect, a sandy soil, will thus produce conditions which may differ greatly from those of a place—perhaps at no great distance—situated on a wind-swept northern slope with a cold clay soil. The character of the climate of a country or district influences, as everyone knows, both the cultivation of the soil and the products which it yields, and thus, indirectly as well as directly, exercises a profound effect upon Man.

In considering the climate of the county of London we must bear in mind that it is not a maritime county like Essex, and so has not the modifying influence of the sea. It will also be well to remember that in point of size, London is the smallest of our English counties, and so we must not expect to find the variations in its climate so noticeable as those in Kent or Essex.

It is of the greatest importance to have accurate information as to the prevailing winds, the temperature, and the rainfall of a district, for the climate of a county has considerable influence on its productions, its trades and industries, and its commerce. Our knowledge of the weather is much more definite than it was formerly, and every day our newspapers contain a great deal of information on this subject. In London the Meteorological Society collects information from all parts of the British Isles relating to the temperature of the air, the hours of sunshine, the rainfall, and the direction of the winds.

The British Isles have been divided for these purposes into various districts, and day by day the newspapers publish the forecasts issued by the Meteorological Society of the probable weather in these districts for the twenty-four hours next ensuing, ending midnight. Thus, for September 30, 1909, the following was the forecast for London, which is placed in the south-east England district: "Calms and very light variable breezes; north-easterly to north-westerly; dull to fair or fine; local rain and mist; cool." Warnings are also issued when necessary, so that the districts may be prepared for any rough weather that may be expected. Besides this information, some of the newspapers print maps and charts to convey the weather intelligence in a more graphic manner.

A glance at a map of the World will show that as the British Isles are in the same latitude as Central Russia, Southern Siberia, Kamtchatka, and Labrador, they get the same amount of heat from the sun and the same duration of day and night, summer and winter; but the direction of the prevailing winds renders available throughout the year much of the heat which the sun has radiated on more southern regions.

The prevailing winds of London, like those of the British Isles generally, are south-westerly. Indeed the wind blows from the south-west for a greater number of days in each month than from all other directions together. A knowledge of this fact helps us to understand that the west end of London is the least smoky and therefore the best quarter for residence. For a short period of the year, London suffers from the east wind, and during its prevalence in March and April, the air is dry and catarrhal complaints are common.

London is inconvenienced during two or three months in the year by thick fogs which hang over the city like a black pall. Their density is mainly due to coal-smoke, and many efforts have been made to remedy this scourge. Fogs can nowhere be avoided in the London area, though they are less dense at Hampstead and Highgate in the north, and at Streatham in the south, than at Whitechapel or Rotherhithe. There is even a Smoke Abatement Society, but up to the present there has been little or no alleviation of the annoyance. Besides the health point of view, there is the commercial aspect of these London fogs, for the dislocation of traffic and business is enormous, while the frequent resort to artificial light results in a great expenditure of money by tradesmen and others. In London, the yellow fogs are known as " London's Peculiar," and Dickens often refers in his works to this name. London has also its mists, and these occur throughout the year. To the artistic eye, a London mist is really beautiful and Whistler has taught people to appreciate them in relation to the river by his " Nocturnes."

The warmest month in London is generally July, when an average temperature of 64° Fahr. prevails, and January, the coldest month, has an average temperature of 39°. The mean temperature of England was 48·7° in 1906, and that of London was 50·7°. With regard to the hours of bright sunshine, we find that while London in the same year had 1734·5 hours out of a possible total of 4459, the average for all England was 1535·5, so that London was much above the country as a whole.

The statistics with regard to the rainfall are arranged in an annual known as *British Rainfall*, and from it we can find exactly recorded the number of inches of rain that fall at about 4000 stations throughout the British Isles.   In the County of London there are many observers who keep one or more rain-gauges and enter the result in a register.   Every year these facts are tabulated for that station, and then forwarded to the editor of *British Rainfall.*   In the County of London there are two very important stations for recording meteorological statistics.   The first is Greenwich Observatory, which is a Government establishment, and the second is at Camden Square, where Dr H. R. Mill, the editor of *British Rainfall,* has a station.

Now let us look at the rainfall statistics at Greenwich for 1907.   During that year there were 163 rainy days with a total rainfall of 22·25 inches.   The two wettest months were April and October, each having a rainfall of over three inches, while March, July, and September were the driest with a rainfall of less than one inch. If we turn to the records at Camden Square, which is in the north of London, we find that rain fell for 418·8 hours on 175 days, with a total rainfall of 23·01 inches, which is rather higher than that of Greenwich.   The average rainfall of 50 years at Camden Square is 25·07 inches; the lowest rainfall of 17·69 inches was in 1898, and the highest of 38·00 inches was in 1903.   For purposes of comparison it may be mentioned that the rainy days for England and Wales were 203 in 1907, and for the same year the rainfall was 36·11 inches. It will thus be seen that London is far below the rest

Greenwich Observatory

of the country. This is largely due to the fact that, as we have seen, the rainfall of England generally decreases as we travel from west to east.

Now to summarise the main facts with regard to the

climate of London, we may say that it is generally dry, with a rainfall far below the average of England and Wales. The climate is healthy and the prevailing winds are from the south-west. The drawbacks to the climate are the fogs of November and December, and the biting east winds of the spring. Perhaps the best testimony to the healthiness of London's climate may be gathered from the vital statistics for 1908. In that year the death-rate per 1000 was only 13·8 against 15·2 for the whole country.

Greenwich Observatory is world-famous. It was founded by Charles II in 1675, and designed by Sir Christopher Wren. For the purposes of navigation the staple work of the Observatory has always been the observation of positions of the moon and fixed stars, to which has been added the care of the chronometers used in the Royal Navy. The meridian marked out by the Transit Circle, the instrument with which these observations are made, is the zero of longitude used by most of the nations of the world, and the mean time of the meridian of Greenwich is the legal standard time for Great Britain. At 10 o'clock and 1 o'clock each day, the accurate Greenwich time is telegraphed to the General Post Office for distribution over the whole country. The observations of positions of the heavenly bodies have been supplemented by the addition of branches dealing with meteorology, magnetism, the observation of the solar surface, and celestial photography. The photographs of comets, nebulae, and the small satellites of planets, taken with the 30-inch Reflector, compare favourably with those taken

The 30-inch Reflector at Greenwich

in other parts of the world under better climatic conditions. The Astronomer Royal lives at Flamsteed House, which forms the main part of Wren's building. The curiously-shaped domes covering the telescopes, the largest of which is a refracting telescope with an object glass of 28 inches diameter, form striking features in the landscape, and the public clock and standard measures of length in the outer wall of the Observatory are always objects of interest to visitors.

12. People—Race. Dialect. Settlements. Population.

London is the most modern of our counties, and yet as a British city it has behind it a history of nearly two thousand years. Alone among all our British towns, it has been a city of world importance through ten centuries, and at the present time it is the most cosmopolitan centre of the whole world. In some of our English counties, such as Cornwall and Somerset, we find distinct traces of the speech and characteristics of the former inhabitants. This is also true of counties nearer London, such as Norfolk and Suffolk, where the Anglo-Danish influence is marked. But in London we have practically no definite survival of the original races which lived in the city; for the Londoner of to-day is either a recent immigrant from the country or from abroad, retaining his provincial or foreign characteristics, or else he is a hybrid of the most intricate ancestry.

It thus will not be necessary to dwell at length on
the various races that have lived in London.   After its
settlement by the Celts, there is no doubt that the
Roman influence was of great importance and that the
natives were Romanised in many ways, as will be gathered
from other chapters in this book.   When the Romans
withdrew, the Saxons destroyed the city, and for a time
it was a desolation.   The Roman villas, baths, bridges,
roads, temples, and statuary were either destroyed or
allowed to fall into decay.   At length, however, London
again became famous as the capital of the Anglo-Saxon
Kingdom of Essex, and continued to increase in size and
importance.   We may assign the renascence of London
to Alfred, who repaired the buildings and rebuilt the
walls.   The building of St Paul's Minster in the tenth,
and of Westminster Abbey in the eleventh, century settled
the question that London was to be a great ecclesiastical
centre of the English people.

During the Saxon settlement of London there were
frequent incursions of the Danes, and they have left their
mark on London, for several of the city churches retain
their dedication to saints of Danish origin.   Thus we
have St Clement Danes church in the west, and
St Magnus and St Olave churches in the east.

Perhaps the greatest change in London was effected
in 1066, when the descendants of the Vikings, or
Northmen, who had settled in Normandy conquered
our land.   Then it was that William the Norman
imposed his will on our people, and made London his
capital.   From the time when William built the White

Tower London has been without a rival, and has drawn
to itself people from all parts of the British Isles and
from all quarters of the globe.

Since the eleventh century no hostile settlement has
been made in London, but there has been a steady influx

Italian Quarter, Hatton Garden

of immigrants, who in many ways have added to the
prosperity of London. For we must remember that
London has nearly always been kind to aliens, especially
to refugees from their own land, whether from political or
religious motives. Flemings were brought over by some
of our Anglo-Norman kings, and Germans from some of

the Hanse towns in Plantagenet times became numerous in London. Then, too, in the sixteenth, seventeenth, and eighteenth centuries, the Protestant Huguenots were driven from France, and found a safe asylum in Spital-fields, Bethnal Green, and other parts, where some of their descendants still thrive. The Jews at various periods have made notable and valuable additions to the popu-lation of London, and have shown themselves supreme in the financial and other departments of the life of the metropolis. Italians have settled largely in the district known as Hatton Garden, while the French are very numerous in the Soho area.

During the later part of the nineteenth century there was a steady immigration of Germans, Poles, and Russians, who settled mainly in the East End of London. These aliens have not been an unmixed blessing, and an Act was recently passed to restrict their landing. The Act, however, is almost a dead letter, and the stream of undesirable aliens continues to flow, and now whole quarters in the borough of Stepney are practically in-habited by these people, who sell their labour at a very cheap rate. Their number is so large that it is now necessary for policemen and other officials to learn Yiddish that they may deal with these settlers more effectively.

We may now pass to the question of dialect, and here our remarks as to the race of Londoners also apply. Owing to the influx of people from all parts of the British Islands, Londoners of to-day have no definite dialect as we should find in Yorkshire, or Cornwall, or

Somerset. It has been well remarked that one of the most certain means of ascertaining the character of a people is afforded by their colloquial idioms. It would be very difficult to apply this principle to the speech of Londoners, which probably includes the worst as well as the best characteristics of our English language. Even in fashionable circles in London we should find the conversation somewhat slangy, and tending to the dropping of the final "g's" of words. Among the toilers of London the speech is loud, and the initial "h" of words is seldom heard. The dialect of the Cockney has passed into a proverb, and has not only worked havoc in London, but has invaded the districts adjacent to the City, to the great detriment of our English language. Here it may be mentioned that the name Cockney is strictly speaking applied to people born within sound of Bow Bells; and perhaps no novelist has used the Cockney speech so largely as Dickens. In his early days it was common for Londoners to convert the letter "w" into "v," and *vice versa*, but this peculiarity does not now exist.

With these few remarks on the race and dialect of Londoners, let us turn our attention to the population of London as it was in 1911, when the last census was taken. For the statistics relating to the population we have no exact information till 1801, when the first census was commenced. From that date onwards, there has been a numbering of the people every ten years.

In 1801 the population of London was 959,310 and in 1911 it was 4,522,961. This means that the population has increased nearly five-fold in the century.

During the last 20 years the increase has been nearly 300,000, but the census of 1911 shows a decrease of 13,306. This enormous population of London is greater than that of either Scotland or Ireland, and exceeds that of fifteen European countries. It forms about one-eighth of the entire population of England and Wales; and while the average number of people to a square mile in England and Wales is 618, it is no less than 38,690 in London.

The population of London north of the Thames is 2,678,651; that on the south side of the river 1,844,310. There are 2,395,804 females, and 2,127,157 males in London's population of 1911. In 1901[1] the people lived in 1,019,546 separate tenements, of which the greater number, or 672,030, had less than five rooms. It was found that there were 40,762 single-room tenements, each having more than two persons, and 1602 single-room tenements with more than six persons in each.

The census figures are interesting in many ways. Thus we find that in 1901 there were 46,646 paupers in London's workhouses, and 4167 prisoners in the gaols. The military barracks of the metropolis held 10,058 people, and there were 10,675 inmates in the various hospitals. To give some idea of the longevity of Londoners, it is interesting to record that 161 people were between the ages of 95 and 100, while 24 people had exceeded 100 years.

The blind people of London in 1901 numbered 3556, and these were largely employed in making articles of willow and cane, as brush-makers, and as musicians. The

---

[1] Details of the 1911 Census are not yet published.

deaf and dumb were 2057 in number, and they worked as tailors, boot and shoe makers, and dressmakers.

Now we come to the census tables (1901) which give the place of birth of the people. We learn that 3,016,580 were born within the county of London, 35,421 in Wales, 56,605 in Scotland, 60,211 in Ireland, and 33,350 in British Colonies. Persons of foreign birth numbered 161,222, and were mainly natives of Russia, Germany, France, Italy, and the United States. The London borough having the largest foreign population is Stepney, where no less than 54,310 aliens were residing in 1901. The boroughs of Westminster, St Pancras, Holborn, and St Marylebone have also a large population of foreigners.

As this volume deals with the western portion of the county of London, we may close this chapter with a few figures giving some comparisons with regard to the populations of the various boroughs. Of the total population of London in 1911, 2,457,533 are in the eastern portion, against 2,065,428 in the western portion. Of all the London boroughs, Islington in the east has the largest population of 327,423 people, while the City of London has the smallest of 19,657 people. Of course the latter is the night population; in the day-time the city population would be ten times as great. Some of the boroughs in the eastern portion are densely populated, and we find that Southwark, Shoreditch, Bethnal Green, and Islington have each 170 or more people to the acre. Woolwich and Lewisham have the least crowded populations, for the former borough has only 14, and the latter only 22 to the acre.

13.　Industries and Manufactures.

London has long been celebrated for its manufactures as well as for its commerce. So early as the reign of Henry I the English goldsmiths had become so eminent for working the precious metals that they were frequently employed by foreign princes. The manufacturers of London in that reign were so numerous as to be formed into fraternities, or gilds. In 1556, a manufactory for the finer sorts of glass was established at Crutched Friars; and flint-glass, equal to that of Venice, was made at the same time at the Savoy. Coaches were introduced in 1564, and in less than 20 years they became an extensive article of manufacture in London.

The making of "earthen furnaces, earthen fire-pots, and earthen ovens transportable" began in Elizabeth's reign, when an Englishman named Dyer brought the art from Spain. The same man was sent at the expense of the City of London to Persia, and he brought home the art of dyeing and weaving carpets. In 1577, pocket watches were introduced into England from Nuremberg, and almost immediately the manufacture of them was begun in London. In the reign of Charles I, saltpetre was made in such quantities in London as not only to supply the whole of England, but the greater part of Europe.

The Huguenots and other refugees from Europe brought their skill and instructed the people among whom they settled how to manufacture many articles

Old Silk-weavers' Houses in Church Street, Spitalfields

(*Showing wide attic windows*)

that had been previously imported. Wandsworth became a busy little manufacturing town in 1573, when a colony of Huguenots introduced the hat manufactory; and it is said that this was the only place where the Cardinals of Rome could obtain a supply of their hats.

The silk manufacture was first established at Spitalfields by the expelled French Protestants, after the revocation of the Edict of Nantes in 1685. For a long time silk-weaving was a most flourishing manufacture, and although it has greatly declined, there are still descendants of the old French Huguenots who live in Spitalfields and Bethnal Green. Foreign names are visible on the shop fronts, and some of the weavers still work in glazed attics such as their forefathers used in France. It is related that the Pope in 1870 wished to procure a silk vestment woven all in one piece. Search was made in France and Italy for a man who could do this, but without success. At last he was found in Spitalfields, and, curiously, the weaver thus discovered was a direct descendant of one of the Huguenot refugees who had left France two hundred years before.

In the seventeenth century, Lambeth was a manufacturing centre, and foreign workmen taught English people the art of making plate glass, delft ware, and earthenware. This last manufacture is still one of its principal industries, and Doulton ware is celebrated all over our country. It may also be mentioned here that Bow and Chelsea were justly celebrated at one time for their china, which is now eagerly sought for by collectors.

One of the most important industries in London is

brewing, and it is interesting to know that the recipe for brewing was brought to London by some Dutch from Holland and Flanders.   London stout and London ale have long been famous, and Stow tells us that in his time there were 26 breweries in the city all " near to the friendly water of the Thames."   Nowadays there are

Pottery-making, Doulton's Works

upwards of 100 breweries in London, and many of them in order to get a good supply of water have sunk wells hundreds of feet deep.   Some of the largest London breweries are in Southwark, Whitechapel, and Holborn. Besides brewing, distilling and sugar-refining are carried on in various parts of London, but the latter industry is rapidly declining.

Tanning and the leather trade have been carried on in Bermondsey and Southwark for hundreds of years, and are still in a flourishing condition. When these trades were first introduced, north-east Surrey had oak-woods, but these have long passed away, and the whole of that district is one of the busiest parts of London. Bermondsey and Southwark have numerous other industries, and soap, candles, and biscuits are largely manufactured.

London has large manufactures of boots and shoes, and ready-made clothing, especially in the East End. A great deal of this work is carried on in the houses of the poor, and gives employment to hundreds of women and children. The making of lucifer matches is also in the same district, and for such work the people are badly paid.

Industries connected with the printing trade, book-binding, and newspapers are of the greatest importance, and London stands at the head of all our towns in this respect. The book-trade at one time centred round Paternoster Row, but of late years many publishers have removed to the West End. Most of the newspapers have their offices in Fleet Street, and in that neighbourhood are the chief printing and bookbinding firms.

London is also one of the great centres of the furniture and cabinet-making trades, and until recent times the principal shops and factories were in the neighbourhood of Shoreditch. These industries are declining, owing to foreign competition, and much of the work has been transferred to the neighbourhood of Tottenham Court Road and Oxford Street.

Clerkenwell is famous for its clock and watch-making, and Hatton Garden is the seat of the jewellery trade and is noted for its dealers in precious stones. Long Acre has a good deal of coach and carriage building; and at Lambeth there are important engineering works.

Although London is a great port, it has not much shipbuilding, and the little that is now done is chiefly confined to the Isle of Dogs, a poor district that has also extensive docks and chemical works. London has always been celebrated for its manufacture of scientific instruments, such as those connected with surgery, optics, and mathematics, but foreign scientific instruments are now competing with the London articles, and this trade is declining.

Woolwich is a district that owes its importance mainly to the Arsenal, which gives employment to as many as 14,000 workmen at a busy time. The artisans are employed in making all kinds of cannon, gun-carriages, shot and shell, rockets, fuses, and torpedoes. This part of Woolwich is thus a crowded hive of workers, and the Arsenal one of the largest and most complete in the world.

14. Trade. The Markets. The Custom House. The Exchanges. The Bank of England. The Royal Mint.

Before locomotion by steam power, the geographical position of London made it the principal port of the whole island; and the introduction of railways and steamships in

the nineteenth century has tended still further to increase
its trade.   From the earliest period of its history, London
has depended on its trade for its supremacy, and Tacitus
speaks of the concourse of foreign merchants in the
London of his time.   Although he does not tell us the
nature of the trade, we know that there was corn to be
shipped from the Thames, as well as tin, and oysters, and
without any doubt among the exports of the Roman
occupation we may reckon slaves.

When London was settled by the East Saxons, trade
came to it again, and under Alfred and his successor it
became the chief port and market-place of our land.
Bede, who died in the eighth century, praises the happy
situation of London on the Thames, and calls it the
emporium of many nations.   Among the names of
London in the seventh and eight centuries, we find
Ceap-stow, Lunden-Wic, Lunden-byrig, and *Lunden tune's
hythe*, and these certainly show the recognised importance
of London as a market and port.

From the beginning of the ninth century the trade of
London is more and more often mentioned.   Then it
is that the sea-faring merchant is rewarded, and the
customs are of such importance as to be worthy of
special regulations.   It is probable that the first port
was at Dowgate on the Walbrook, but as larger ships
came, they moored alongside the Thames at Billingsgate
and Queenhithe.

A blow was struck at the slave-trade in 1008, when
it was decreed that " Christian men, and uncondemned,
be not sold out of the country," and within a few years

such merchandise was not necessary for the growing prosperity of the port. One remarkable fact is that the early commerce and trade of London were mainly in the hands of foreigners. Indeed it has been said that London in those early times was largely a city of foreigners. In order to encourage our own people to trade, Athelstan, early in the tenth century, ordained that a merchant who had made three voyages should be of right a thane.

Among the foreign traders who settled in London were some Germans, who were known to the English as Easterlings. They had their own hall, or Gildhall, which was called the Steelyard and stood on the site of the present Cannon Street Station. These Teutonic or Hanse merchants had a monopoly of all the trade with the nearer countries of Europe, and they flourished in London till the reign of Elizabeth, when they lost all their special privileges.

The Italian money-lenders, known as Lombards, began to settle in London in the thirteenth century. They were mostly wealthy Italians driven from their own country, and owing to the expulsion of the Jews from London, they did a large and profitable business. These Italian money-lenders have left their mark on London trade, and Lombard Street, named after them, is still the seat of our chief banking houses.

The reign of Edward III saw the increase of our trade with the Low Countries, and the settlement of Flemings in London and elsewhere. Flanders became the great market for our wool and so continued down to the time of the Tudors. Elizabeth gave a great stimulus to the trade of London, and it was in her reign that Gresham's Royal Exchange was built. When Antwerp was sacked by the Spaniards in 1576, London took its place as the leading port of Europe. The Hanse merchants lost their privileges by the action of Elizabeth, and at once the Merchant Adventurers took their place and carried on the wool trade with Flanders. It was owing to Elizabeth's wise commercial policy that other companies were formed for the development of trade. The Russian Company brought the furs of Russia to London, besides silks and, later, teas from the East. The Levant Company developed a trade in the Mediterranean, and the East India Company began its work during the last year of Elizabeth's reign. All these trading companies enjoyed monopolies, but they brought increasing prosperity to London.

We can best grasp the present trade of London by looking at a few figures. The whole of the imports of the United Kingdom in 1910 exceeded the value of £678,000,000 and of this London took £228,000,000 or more than one-third. The total exports of the United Kingdom for the same year were over £534,000,000, and London's share of this was about £122,000,000. It will thus be seen that nearly a quarter of the trade of our country is carried on through London; and it will also be noticed that the imports exceed the exports by nearly two to one. London is therefore the chief port for imports, and is exceeded in its exports only by Liverpool.

Now it will be interesting to consider the chief articles that London imports. Nearly all the wool that comes to our country enters the Port of London, and its wool market is attended by buyers from all parts of the world. Most of the tea and coffee consumed in England are brought to London, and the City has practically a monopoly of the fur trade with Canada. London has a large share of the West Indian trade, which includes cocoa and sugar; and about a quarter of the tobacco trade belongs to it. Petroleum from America and Russia, cheese from Canada, and timber from the Baltic ports are almost monopolies of London. It is calculated that half of the wine that comes to England pays duty at the Custom House; and all kinds of French manufactures are sent to London. Among the chief exports of London are cotton goods, metal manufactures, wearing apparel, woollen goods, and machinery.

We will now pass to a consideration of the markets of London. Markets have been in existence in the City for more than a thousand years, and for many centuries the City has been the market authority for London. The City was granted in the reign of Edward III exclusive market rights and privileges within a radius of seven miles, and these rights have from time to time been recognised and confirmed.

It will thus be seen that nearly all the markets in the County of London are under the control of the City Corporation. The Central Markets at Smithfield, the Cattle Market at Islington, the Foreign Cattle Market at Deptford, the Fish Market at Billingsgate, and the

Meat and Poultry Market at Leadenhall, are managed by the City authorities, and are noticed in the volume on East London.

In this volume we need not concern ourselves with the many street markets which supply the poorer classes with food of all kinds, nor with the small markets in

Covent Garden

private hands.   There is one market, however, although not managed by a public body, which merits more than a passing notice.   Covent Garden Market is justly famed for its fruit, flowers, and vegetables, which come from all quarters of the world.   It takes its name from the fact that it is on the site of the garden belonging to the

Convent of Westminster, but it was not till the sixteenth century that it was commonly written Covent, as being derived from the French *couvent*, rather than from the Latin *conventus*.

In course of time this property of the Convent at

Covent Garden Porters

Westminster passed into the possession of the Earl of Bedford, and the present market belongs to a descendant of that nobleman. It seems to have originated in the middle of the seventeenth century in a few stalls and sheds on an open space, but the present market-place was erected in 1830 by the sixth Duke of Bedford. Since that date the market has continued to grow, and the French flower market and the English flower market are some of the later additions. The market days are Tuesdays, Thursdays, and Saturdays, and then the show of flowers is seen to best advantage at an early hour in the morning. The Easter Eve flower market is particularly brilliant, and gives one an idea of the wealth of flowers used in the decorations of the churches of London at Eastertide.

In this volume we can only glance at such institutions as the Custom House, the Exchanges, the Bank of England, and the Royal Mint. Situated as they all are in the eastern portion of London, they are dealt with at greater length in the first volume. The Custom House in Lower Thames Street is the building where the customs are collected for the Port of London. It was erected in 1814 and has a fine river frontage of 488 feet, the quay forming a noble esplanade with a good view up the river.

The Stock Exchange in Capel Court is quite close to the Bank of England. It is known in the City as "The House," and its 3000 members have the right to buy and sell stocks. The Royal Exchange, the third on this site, is one of the best known of the City buildings, and was opened by Queen Victoria in 1844. The eastern part of the Royal Exchange is occupied by "Lloyds," an association of underwriters whose business has largely to do with shipping and insurance. Among the other Exchanges in the City are the Corn Exchange in Mark Lane, the Shipping Exchange in Billiter Street, the Wool

Exchange in Coleman Street, and the Coal Exchange in Lower Thames Street. In each case the name is indicative of the class of business transacted in these important buildings.

The chief banks have their headquarters in Lombard Street, or in its neighbourhood, and branches of them are to be found in all parts of London. When we speak of the banks of London we naturally think of the Bank of England, which was founded by William Paterson, a Scot, in 1694. The Bank of England transacts much business for the Government, and is the only bank in London which has the power of issuing paper money.

The Royal Mint on Tower Hill is connected in some ways with the Bank of England. All the gold bullion from the Bank of England is here made into sovereigns and half-sovereigns, and in addition all our bronze and silver coins are produced here. The value of our Imperial coinage issued by the Mint in one year amounts to as much as £7,000,000.

15. History.

There is no authentic evidence to show at what period the Britons settled in the district we now call London. We have already referred to the origin of the name, and also to the position of the city in the days of the Britons. Here we need only remark that it was then probably little more than a collection of huts on a dry spot in the midst of a marsh, and surrounded by an

earthwork and ditch. It is not possible to say what length
of time elapsed between the foundation of British London
and its settlement by the Romans. We do know, however, that after the Roman conquest London rapidly
grew in importance, and in A.D. 61 it is spoken of as
a place noted for its concourse of merchants.

The early Roman city on this site was called Augusta,
and was founded in the reign of Nero, A.D. 62. As a
Roman city it did not rank in importance with either
Eboracum (York) or Verulamium (St Albans), and it was
never regarded as the capital of Roman Britain. Under
the Romans, the city extended from the site of the present
Tower of London on the east to Newgate on the west,
and inland from the marshy banks of the Thames to some
swampy land known as Moorfields. The Romans left
their mark on London, and the wall with its gates, and
the bridge over the Thames which they built, are all
referred to in other chapters.

The actual historical references to London during
the Roman period are very few. In the year A.D. 61
Boadicea, the British queen, attacked the town, and
Suetonius and his troops were forced to evacuate it.
For more than two centuries after its capture by the
British forces we have no mention of London by any
historian. There is every reason to believe that the
Romans recaptured it, and that its prosperity increased.
Towards the end of the third century, Carausius, the
Roman commander in Britain, proclaimed himself em-
peror, and struck in London gold coins bearing his portrait
and name. Before long, however, he was murdered by

Allectus, who assumed the imperial title but was defeated
and slain at Southwark by a general whom the Emperor
of Rome had sent against him. Henceforward the history
of London becomes fragmentary, and about 410 the
Roman soldiers were withdrawn, and Augusta was
forgotten for a period of nearly two hundred years. We
may say that Augusta had perished, and when the City
comes again into the light of history, it is under its more
ancient name—London.

This silence of history for two centuries is very remarkable, for we hear of the Saxon conquest of Pevensey,
Bath, and Gloucester, but the story of London is quite
lost to us. In the year 604 A.D. we find the city in
the possession of the East Saxons, and new names are
given to the old Roman roads, the gates, the rivers,
and the hills. Everything is changed, and the power
of Rome over London has vanished. London became
the capital of the kingdom of the East Saxons, and
continued to increase in size and importance. As early
as the beginning of the seventh century the influence
of Christianity made itself felt in London. Ethelbert,
King of Kent, had been converted, and as overlord of all
nations south of the Humber, he had a sincere desire that
the East Saxons should become Christians like the people
of Kent. He therefore decreed that the people of London
should put away the worship of Thor, Odin, and other
gods of the north. Evidently he was obeyed, and Mellitus
was consecrated the first bishop of London in 604. Bede
tells us that Ethelbert built the first church of St Paul in
London, and the site of the present Westminster Abbey

was also occupied by a Christian church. Many other churches were built in various parts of London, and were dedicated to national and local saints, such as St Dunstan, St Botolph, St Osyth, and St Swithin.

From the time of its conversion, London steadily grew, and learning and culture came from over the sea to its people. Monasteries were founded, and monks from the continent came to fill them. The arts of architecture, painting, and music developed, and a brighter and better life for the dwellers in London was the result. The Christian Saxons of the seventh and eighth centuries were quite unlike the pirates and plunderers of earlier days. Peace had settled on the land, and, as a result, commerce brought riches to London. Early in the ninth century, however, a new enemy appeared at the gates of London, and for a long time the Danes harried the city. At length, in 839, they captured it, and made it their headquarters. The work of the Danes was comparatively easy, for as the English were not distinguished as builders, they had not strengthened the fortifications of London.

At the time of the occupation of London by the Danes, Alfred was king, and he made it his aim to recapture the city. He recognised the value of London as a possession, and in 884 the city fell into his hands. The name of Alfred is imperishably connected with London, and one of our historians goes so far as to say that he gave us London. At any rate in the year 886, Alfred determined to rebuild and strengthen the city. The *English Chronicle* says, " Alfred honourably rebuilt the city of London and made it again habitable." Never again did the Danes conquer London, and from that date London has continued to increase in wealth and prosperity.

The Danish occupation of London may be traced in some of the names of its churches and streets. Not only, as we have seen, were churches dedicated to St Magnus, St Olave, and St Clement, but we are reminded of the Danish settlement by Tooley Street, which is a corruption of St Olaf's Street, and Gutter Lane, off Cheapside, which is said to be the modern form of Guthrum's Lane.

The last event of importance in Old London was the building of Westminster Abbey by Edward the Confessor. This was not the Abbey that we see to-day, but it was a church built on what was then the swampy island of Thorney, and around it grew up the city of Westminster.

During the early period of our history, London fought an uphill fight with Winchester for the position of chief city of England, and it was not till the reign of Edward the Confessor that it became the recognised capital of our country. This position was still further secured when William was crowned in Westminster Abbey on Christmas Day, 1066 ; and in return for the submission of the citizens, William granted them a charter, which was of immense importance to them. Having gained possession of London, William proceeded to fortify the city. He enclosed a space of about twelve acres in the east of London, and gave orders to build the Tower. Gundulf, Bishop of Rochester, was the architect, and the White

Tower as we see it to-day is the most important remnant of Norman London.

In the year 1100, Henry I granted the citizens of London a second charter, and from the advantages it conferred we may measure the growing importance of the city. We have the first record of the Mayor of London in 1190, when Henry Fitzailwin was elected

The White Tower

to that high position, which was held by him for twenty-four years. London was recognised as a *communa*, or fully organised corporation in 1191, and when King John was quarrelling with the barons, the Londoners took the side of the latter. The Magna Carta specially secured to London its rights and customs as obtained in previous charters, and gave it the power to elect its mayor annually. It also ordained that the Mayor of London should aid the twenty-four lay peers to compel the king, by force of arms, to keep the Charter.

The reign of Edward III was remarkable in the history of London, for royal charters were granted to some of the craft-gilds, and henceforward the Livery Companies had a direct share in the government of the city. The power of these companies was enormous, but it was mainly owing to them that London became the first industrial and commercial city in the kingdom.

The Londoners of the time of Richard II showed their power, for they refused the king a loan, and though he deprived them of their charters, it was not long before they were restored. When Wat the Tiler and his followers entered London in 1381, John of Gaunt's palace in the Savoy was burnt, and some of the prisons were opened. While Richard himself was meeting some of the insurgents at Mile End, a large body of them broke into the Tower and murdered the Chancellor and the Treasurer, and other officials. Next day, Richard ventured again to meet them. Wat the Tiler, their spokesman, was so insolent that Walworth, the Mayor of London, cut him down. The angry multitude were dispersed after the king had promised that their grievances should be remedied.

King Henry V entered London in triumph in 1415, after his great victory at Agincourt. Richard Whittington, the mayor, entertained that king at a banquet at his own private house, and the citizens were most enthusiastic in their reception of the victorious monarch. The king attended a thanksgiving service at St Paul's, and then retired to his palace at Westminster.

The reign of Henry VI is memorable in the annals of London for the capture of the city by Jack Cade in 1450. With a large force of Kentishmen, Cade entered London without meeting with any resistance, and, riding up to London Stone, he struck it with his sword, exclaiming, "Now is Mortimer lord of this city." For three days his followers plundered and burnt, until by the exertions of the mayor and aldermen, the rebels, who had retired to Southwark, were shut out of the city.

When the Wars of the Roses began, the Londoners espoused the cause of the Yorkists, and when Edward of York appeared before the gates of London the citizens received him with acclamation. A little later, the people met one Sunday in an open space near Clerkenwell, and with the familiar shout of "Yea! Yea!" they chose Edward IV to be their king, who to the day of his death was popular with the citizens of London. The story of London under the Plantagenets ends with the reign of Richard III, who like his brother, Edward IV, was invited by the mayor and chief citizens to be the King of England.

During the Tudor period London continued to grow in importance. Reference will be made in another chapter to the dissolution of the religious houses in the reign of Henry VIII, but here we may note that London was the chief scene of the burning of "heretics" at Smithfield in the reign of Mary. When Elizabeth was on the throne, the capital showed its patriotism by its liberal contributions of men, money, and ships for the purpose of resisting the threatened attack of the Armada.

Under the Stuarts the history of London assumes even greater importance. Owing to the exactions of the Star Chamber it sided with the Roundheads, and became the centre of Presbyterianism and of opposition to the king. In 1648 the city was occupied by the Cromwellian troops, and in the following year Charles I was beheaded at Whitehall. Cromwell was proclaimed Lord Protector of England in 1653, and after his death London was occupied by Monk's troops. The year 1660 witnessed the Restoration of Charles II, who was received back to his kingdom with the greatest satisfaction by Londoners.

The reign of Charles II is memorable in the history of London for two great events. The Plague of 1665 turned the capital into a city of mourning and desolation, and it is calculated that about 100,000 Londoners died of that fell disease. The following year witnessed another dire calamity, for the Great Fire destroyed no less than 13,000 houses and 89 churches. This disaster, however, proved beneficial to London, for it was rebuilt in an improved form. Its streets were widened, and the wooden houses gave place to buildings of stone and brick. The Monument on Fish Street Hill was finished in 1677 as a memorial of the Great Fire. It is 202 feet high, and nearly 202 feet distant from the spot where the fire first broke out on September 2, 1666.

It was not till the reign of Queen Anne that London began to assume anything like its present appearance. It was during her reign that the results were evident of Wren's rebuilding the Cathedral of St Paul's, and the

The Monument

many churches in the city. During the eighteenth century London increased in size and population, and during the latter part of that period some of its handsomest streets were made, and some of its finest buildings erected. Great injury was inflicted on the city by the Gordon Riots of 1780. Lord George Gordon put himself at the head of 40,000 rioters with the cry of "No Popery!" Some of the prisons were destroyed, the prisoners released, and mansions were burned or pillaged. The rioters were not subdued, till nearly three hundred had been killed and Lord George Gordon had been sent to the Tower.

An important event in the social life of the city took place in 1803, when the lighting of the streets with gas was begun. Pall Mall was the first street so lighted, and Bishopsgate Street followed in a short time. The story of London throughout the nineteenth century is one of remarkable growth and expansion, and London has now made good its claim to be not only the largest, but also one of the finest cities in the world. The Metropolitan Board of Works was formed in 1855 to look after the sanitary arrangements of London, and in 1889 this body gave place to the London County Council, whose aim must be, in the words of Lord Rosebery, "to make it more and more worthy of its central position, of its great history, and of its immeasurable destinies."

16. Antiquities—Prehistoric, Roman Saxon.

The conditions of man's existence on the earth during the early stages of his history are shrouded in obscurity. The earliest evidence of the presence of man in London is not by written records, but by implements of chipped flint. The very earliest of these implements belong to a time when our country was joined to the continent, and their age must be reckoned by thousands, if not by tens of thousands of years.

Antiquaries have divided this early period of our country's history into the Stone Age, the Bronze Age, and the Early Iron Age. It must not be thought that, in the Bronze Age, stone had been discarded for many purposes, or that bronze was no longer in common use after the discovery of iron. On the contrary, each material survived long into the succeeding period; but this classification is convenient because it shows the material chiefly in use. These three Ages or Periods cover a vast extent of time, but it is not necessary to say how many years are included in each of them, for we cannot be certain when one age ended and the next began.

The Stone Age has been further divided into the Palaeolithic or older section, in which the flint implements were formed simply by chipping, and the Neolithic, or newer section, in which they were more carefully worked, and even polished. There is reason to suppose that an immense period of time separated the two Ages.

Palaeolithic implements had no doubt been found before it was recognised that they belonged to a remote past; but the first recorded discovery of the kind was made in London towards the end of the seventeenth century. A fine

Palaeolithic Flint Implement found in Gray's Inn Road

pear-shaped implement was found with an elephant's tooth near Gray's Inn Road, and was described, wrongly of course, as a British weapon.

As London lies in the valleys of the Thames and Lea,

it is not at all remarkable that so many traces of man are found within the borders of the county. Both river-valleys have yielded many examples of flint weapons and implements of the Palaeolithic Age. In the bed of the Thames a good many specimens have been found at various times, and they are frequently brought up in the course of dredging operations.

A palaeolithic floor has been traced at Stoke Newington, Stamford Hill, Kingsland, and in the City of London, and many specimens of flint implements from these localities are to be seen in the British Museum. Some flakes found at Stoke Newington Common and struck off by palaeolithic man from the same core are shown fitted together again, as evidence that the manufacture of flint implements took place on this site.

In the Guildhall Museum there is a fine collection of London antiquities, and among them are implements from the river-drift gravels that have been discovered at Bishopsgate, Wandsworth, Clapton, and other places. The best way of learning about the stone implements is to visit either the British Museum or the Guildhall Museum, and there, with a good guide, carefully to examine them.

When the Neolithic Age began, Great Britain had ceased to be a part of the continent and was an island. The climate had become more temperate and rather moist, while such animals as the mammoth had become extinct. Man had now learnt to train animals for domestic use; and he cultivated cereals for food, and various plants to provide materials for woven garments. He used the bow as his weapon, and he had also developed the art of making pottery.

In the Neolithic Age, the implements and weapons were commonly hafted and made in a greater variety of forms; and by the addition of grinding and polishing it was found possible to use other hard stones in addition to flint. While the grinding and polishing of stones may be considered the special characteristic of this period, it must not be supposed that this was always the case. For instance, a large and important class of implements and weapons, such as knives, scrapers, and arrow-heads were but rarely ground or polished, while even axes of fine workmanship were sometimes finished by simply chipping them.

Among the implements found in London belonging to the Neolithic Age may be mentioned celts from Paddington and Southwark; scrapers from London Wall and Battersea; a flint knife from Addison Road, Kensington; and flint flakes from Fulham and Hammersmith.

As the Neolithic Age advanced, man gradually learnt the use of metal, and from this important step in human progress we find traces of his rapid advance in all directions. The period from the beginning of metallurgy down to the dawn of written history is generally divided into two parts:—an earlier or Bronze Age, and a later Age of Iron. Among the antiquities found in London belonging to the Bronze Age may be mentioned a bronze sword of leaf form from the Victoria Embankment, a bronze dagger-blade from the Thames, an implement of red deer-horn from Philpot Lane, and a fragment of pottery ornamented with herring-bone pattern from Hammersmith. Besides these antiquities, socketed celts, winged celts, and palstaves have been found in various parts of the county of London.

Under the portico of the British Museum is a " dug-out " boat which probably dates from the Bronze Age. It belongs to a common type, formed out of a tree-trunk split lengthwise, the work of hollowing the interior being performed by tools of stone or bronze, and possibly by fire. A boat of this kind was found during excavations for the Royal Albert Dock at North Woolwich in 1878. The oak trunk was carefully worked, the bottom and sides being flat and rectangular, but there are no signs of keel, stretchers, or rowlocks.

When the knowledge of iron and the valuable properties it possessed became known in Britain, it is probable that the new metal supplanted bronze in the manufacture of such implements as sword-blades, daggers, and knives. Iron is believed to have been brought into our country by the Brythons, a branch of the Celtic family, and from them our island received its name of Britain. During this age, man in Britain made great progress in culture and the arts, and the antiquities of this period bring us down to the conquest of Britain by the Romans. Many articles of great interest belonging to the Early Iron Age have been found in London, and they include all kinds of personal ornaments, such as fibulae, hair-pins, and rings, various implements and weapons of iron, such as knives, swords, and daggers, and many specimens of pottery, such as vases and urns.

A bronze shield decorated with red enamel was found in the Thames near Battersea, and an examination of this impresses one with the beauty of the curves and the well-balanced proportions of the various parts. The

Enamelled Bronze Shield
(*From the Thames at Battersea*)

boss of another shield was also found in the Thames near Wandsworth, and the decorations on it are produced by means of nearly complete circles, enclosing leaf-like thickenings. Both these examples represent the very best work of the Early Iron Age, but there have been found other articles of great merit. A bronze brooch from the Thames, and a scabbard with mock spirals, also from the Thames at Wandsworth, are extremely interesting specimens of this period.

Before we leave this early Iron Period, it may be well to ask what monuments are left in London to remind us of the early Celtic people in our land. Of prehistoric monuments in and around London there are only two, and they are the Hampstead Barrow and the River Walls. The Hampstead Barrow, locally called Boadicea's Grave, is of doubtful origin. It was opened in 1894, but nothing was found in it, so that it may be only a boundary hillock. With regard to the river walls, which may be seen from Barking on to Southend, the tradition is that they were first built by the Britons. There is no date for them; but although they have been often repaired, they are practically the same as when first constructed. We found in the first chapter that the word London may remind us of the Britons, the early Celtic people in our land. If the word London is derived from Caer-Lud, after a King Lud of Celtic history, then we have in the present name of the county a direct link with this early period of our history.

The Roman remains found in the county of London are very numerous, and from them we are able to picture to ourselves the life of its people two thousand years ago. It has been said that the two chief events in the history of Roman London are the building of the bridge and the building of the walls, but as both of these are considered in other chapters, we need make no further reference to them here.

Roman London is now a buried city, and all the remains of it have been found many feet below the level of the present streets. The smaller antiquities include personal ornaments, such as bronze fibulae, rings, brooches, hair-pins, ear-rings and gems, and many of these articles have been discovered in various parts of the City of London, especially in London Wall and Aldgate. Domestic utensils and appliances are frequently unearthed, and also such things as Roman balances, bowls of bronze, knives, and spoons, which have been found near the Mansion House, in Queen Victoria Street, and in London Wall. Iron lamps and lamps of glazed ware have been unearthed in Lothbury and Broad Street, and all kinds of tools, such as chisels, axes, adzes and piercers, have been discovered in London Wall and at Wapping. Very numerous discoveries have been made of metal objects such as bells, chains, and horse-furniture at Southwark, Austin Friars, and Walbrook.

Among the more interesting Roman remains in London may be mentioned the figures and statuettes in metal, clay, and terra-cotta. In the Guildhall Museum there is a very fine collection of these objects, and statuettes of Apollo, Hercules, Juno, Mercury, Mars, and Venus are quite numerous. From them we get a good idea of Roman art, and also a realisation of the chief Roman deities.

From the Roman remains in London we can also picture to ourselves the internal decoration of the houses

of the citizens.   No less than 40 mosaic pavements, either
complete or incomplete, have come to light, and of these
the most beautiful and perfect is the mosaic pavement
found in Bucklersbury in 1869.   It is composed of red,
white, grey, and black tesserae ; the lower and main
portion is in the form of a parallelogram, while the
upper part is semi-circular.   The central device is a
floral design, surrounded by a cabled band, and enclosed
within two squares of ornament placed at different angles,
having a floral device at each corner.   Between the upper
and lower portions is a broad band of floral scrolls.   The
upper semi-circular portion has a fan-shaped device in the
centre, above which is a pattern of scale ornament, and
the border consists of a knotted band.   The whole design
of this fine mosaic pavement is enclosed in a border of
red tesserae.

Roman glass and pottery have been found very ex-
tensively in various parts of London.   The glass urns and
bottles, as well as bowls, cups, and dishes of glass are
of considerable value on account of their form and colour,
and the pottery of various wares makes perhaps the most
striking appeal to a visitor to the Guildhall Museum.
Here may be seen urns, vases, bowls, and cups of Samian
and Upchurch wares, which have been dug out in Aldgate,
London Wall, Bishopsgate, Lombard Street, and other
places.   In 1677, on the site of the present St Paul's, a
Roman kiln was discovered, where coarse pottery, as well
as all kinds of tiles, were manufactured.

The Romans had burial-places and tombs in Bow Lane,
Camomile Street, Cornhill, and St Paul's Churchyard,

and sarcophagi of stone and marble have been found at Clapton and the Minories. Leaden coffins were discovered at Old Ford in 1844, and at Bethnal Green in 1862, and monumental stones and slabs are sometimes unearthed. In 1911, a Roman galley was found when excavating for the foundations of the new County Hall at Lambeth. It is nearly 50 feet long and weighs more

Roman Boat, found near Lambeth, 1911

than six tons. It has been removed to Kensington and will be kept in the London Museum now being formed.

There is one other relic of Roman London to which we must briefly refer. London Stone is built into the south wall of St Swithin's Church in Cannon Street. It is probably an old Roman mile-stone, which may have marked the beginning of the first mile on the Watling Street. London Stone is one of the most valued relics of London, and in the Middle Ages it was very greatly esteemed.

When we pass from the Roman period to the time of the Saxons, we find a positive dearth of antiquities to represent this later age. Besant says "there is nothing,

London Stone, Cannon Street
(*From an old print*)

absolutely not one single stone, to illustrate Saxon London," but he thinks that some of the columns in Westminster Abbey and the Chapel of the Pyx represent the work of the Confessor. In 1774 an earthen vessel containing coins of Edward, Harold, and William was found near St Mary-at-Hill, and an enamelled ouche in gold of the ninth century was discovered near Dowgate Hill. Personal ornaments and requisites of metal, bone, and horn, as well as weapons and tools of the Saxon period, have been dug up in various parts, but they are not of sufficient importance to be specified.

What has survived to remind us of early London is of more importance perhaps than many relics. The names of many streets such as Cheapside, Ludgate, Bishopsgate, Coleman Street, Cornhill, Walbrook, and Gracechurch Street are without doubt of Saxon origin ; and the churches dedicated to St Botolph, St Ethelburga, and All Hallows remind us of some of the Saxon saints. Again, the influence of the Saxons is evident by the usage to-day of such words as alderman and sheriff, and of the name of the meeting-place of the councillors—the Guildhall.

17. Architecture — (a) Ecclesiastical. Medieval Churches. Wren's Churches. Chapels Royal.

When we consider the ecclesiastical architecture of London, we find that it occupies quite a different position from that of our other English counties. In them we know that there are hundreds of old English churches, some dating from Norman and even earlier times, but in London the majority of the churches are modern. This fact, of course, is largely owing to the Great Fire of 1666, which swept away some of the finest churches of the earlier periods.

Before we notice the results of the Great Fire, it will
be well to glance at the ecclesiastical condition of London
before 1666. Perhaps the first thing that strikes us is
the large number of parishes, each with its own church,
within the walls of ancient London. Fitzstephen tells us
that in his time there were 13 large conventual churches,
and 126 lesser parochial churches; and later Stow, the
great historian of London, gives a list of 125 churches,
including St Paul's and Westminster Abbey.

At the present time the City of London has many
fine churches, and from the Surrey side of the Thames
one is struck with the lofty steeples and spires that tower
in their beauty above the warehouses and places of
business. Let us think for a moment what place the
church occupied in the medieval life of London. Then
the resident population of the City was much greater than
it is now, and many of the parishes were very small. At
every street corner rose a church, and the City was filled
with priests, friars, and other ministers of the church.
While some of the churches were small, many were rich
and costly buildings, which had been beautified and
adorned by the loving thought of many generations.

The church then occupied a large part of the daily
life of the people, who were expected to attend service in
their parish churches. The gilds and trade companies
went to church in state. All the people belonged to the
Church and its services were part of their life, as much
as their work, their food, and their rest.

We need not go into details as to the changes that
were brought about by the Reformation, or to the ravages

of the Great Fire.   We may note, however, in passing
that 89 churches were destroyed in 1666, and of them
only 45 were rebuilt.   Among the churches that escaped
the Great Fire were St Giles' Cripplegate, St Helen
Bishopsgate, St Bartholomew's Smithfield, St Katherine
Cree Leadenhall Street, All Hallows Barking, St Olave
Hart Street, St Etheldreda Ely Place, St Ethelburga
Bishopsgate Street, St Andrew Undershaft, and the
Church of the Austin Friars.   Pepys the diarist has the
following observation on the subject of the London
churches destroyed in the Great Fire:—"It is observed
and is true in the late Fire of London, that the fire burned
just as many parish churches as there were hours from the
beginning to the end of the Fire; and next, that there
were just as many churches left standing as there were
taverns left in the rest of the City that was not burned,
being, I think, thirteen in all of each, which is pretty to
observe."

There are, however, remains of early and medieval
churches in London, which are worth considering and
are representative of all styles of English architecture.
The successive periods of English church architecture are
generally given as Norman, Early English, Decorated, and
Perpendicular, and range in date from the eleventh to the
sixteenth century.

The Norman period of church architecture was from
1070 to the later part of the twelfth century; and of
early Norman work London has some good examples in
the chapel of St John in the White Tower, and the
crypt under the church of St Mary-le-Bow, Cheapside.

The choir of St Bartholomew, Smithfield, and some details in the nave of St Saviour's Cathedral, Southwark, illustrate later Norman work.

Towards the end of the twelfth century the round arches and heavy columns of Norman work began gradually to give place to the pointed arch and lighter style of the period of English church architecture known as Early English, which is so conspicuous for its long narrow windows. The choir and eastern chapels of St Saviour's Cathedral and Lambeth Palace Chapel exhibit the Early English style at its best. The transepts and choir of Westminster Abbey are most glorious examples of Early English, and this building illustrates English Gothic architecture in all its phases.

The Early English style flourished from 1154 to 1270 and gave place to the highest development of Gothic— the Decorated, which prevailed throughout the greater part of the fourteenth century, and was particularly characterised by its window tracery. The chapel of St Etheldreda in Ely Place, Holborn, and the lower chapel of St Stephen in the Houses of Parliament, illustrate the Early Decorated, while the windows in the nave of the Austin Friars' Church, and the south transept of St Saviour's Cathedral, give a good idea of the work of the Late Decorated period.

The Perpendicular is the name given to the last period of Gothic architecture in England. It established itself towards the end of the fourteenth century and was in use till about the middle of the sixteenth century. The Perpendicular, which, as its name implies, is remarkable for the perpendicular arrangement of the tracery, and also for the flattened arches and the square arrangement of the mouldings over them, is only seen in England. It is to the Perpendicular period that the old churches situated in the northern and eastern parts of the City chiefly belong. They are St Giles's Cripplegate, St Helen's and St Ethelburga's Bishopsgate, St Andrew Undershaft, St Olave's Hart Street, All Hallows Barking, and St Peter ad Vincula within the Tower.

There is not space to go into details with regard to the old churches of the parishes outside the City of London. As a rule they are not elaborate, and offer few examples of artistic detail. In no way do they compare favourably with those of Essex, Suffolk, or Norfolk, either in point of size or beauty of parts. Plainness and simplicity are the characteristics of these old churches; and the plans are almost always a nave and chancel, a south porch, and a western tower. Their architects had to build with the materials they could command; and as these were different from what are found elsewhere, these churches are unlike those in the City of London, where the most expensive materials were used in the fabrics of the fine churches within the walls.

It will be recognised that the previous remarks apply only to those churches in the county of London prior to the Reformation. From that period to the Great Fire, there was little church-building. But when the Great Fire wrought such havoc it was decided to erect 53 new churches upon the sites of those burnt, or so much damaged as to require rebuilding. Sir Christopher

Wren was the great architect to whom was entrusted this work of church building and restoration, and it is generally acknowledged that the results were excellent. A recent writer says that "Nothing that has been achieved in modern architecture has surpassed the beauty

Sir Christopher Wren

of their steeples, not only from the elegance of each, but for their complete variety, while at the same time in harmony with one another. No two are alike. The view of the City of London from the old Blackfriars Bridge—up to the middle of the last century—must have been scarcely surpassed in any country : and all this was the work of one man !"

St Clement Danes, Strand

Among the best of Wren's churches in the western part of London the following are the most noteworthy :

9—2

St Andrew's Holborn, St Anne's Soho, St Clement Danes in the Strand, and St James's, Piccadilly. Wren's churches are remarkable for the variety and originality shown in their design ; their elegant proportions are noteworthy and satisfactory; and it may safely be said that no subsequent English architect has approached the fine work of Wren.

His work was continued by his pupils, more especially by Nicholas Hawksmoor and James Gibbs. The fine church of St Mary-le-Strand, built at the beginning of the eighteenth century, was designed by Gibbs, although borrowed almost entirely from Wren. The church of St Martin-in-the-Fields is another building of this period, and of the Renaissance style of architecture. Perhaps this is one of the most successful of the designs of Gibbs, forming a composition worthy of Wren himself. The chief features are the great spire and fine portico at the west end.

The church of St George, Bloomsbury, was designed by Hawksmoor. It has a very fine portico, which may have been suggested by that at St Martin-in-the-Fields. St George's is said to be the only church which has a statue upon the steeple. The statue is a full-sized figure of George I clad in a Roman toga, and was the gift of a great admirer of that king. This steeple has been called ridiculous, but the other parts of the church show some really fine features.

It was during the reign of Queen Anne that it was decided to build 50 new churches in London, and of these it may be said generally that they are of the Palladian character and exhibit solidity and grand proportions.   St John's Westminster, was the second of the 50 churches.

St George's Church, Bloomsbury

It was designed by Archer, and is remarkable for its semi-circular arches on the east and west, and for its quartette of belfries, one at each of the four corners. The best known of Queen Anne's 50 new churches is St George's, Hanover Square, which was built in 1724. It has a fine classical portico, with a pediment supported by six Corinthian pillars, and possesses some good sixteenth century windows brought over from Mechlin.

The church architecture of London was largely affected in the nineteenth century by the Oxford Revival, and the Classical style of Wren and his successors gave place to the Gothic. Such architects as Scott, Butterfield, Pearson, Street, and Bodley designed some of the finest modern churches in London. We may specially mention St Giles's Camberwell, St Andrew's Well Street, and St Mary Abbots Kensington by Scott, and All Saints Margaret Street, and St Albans' Holborn by Butterfield, as giving the most satisfactory expression to the eccle-siastical Gothic, and its adaptability for purposes of Christian worship. There is a more recent church than either of those mentioned, which even better exemplifies the beauty of the Gothic style of architecture—the church of the Holy Trinity in Upper Chelsea, designed by Sedding, which gave opportunity for many of the most distinguished artists and craftsmen of the Victorian period to use their varied gifts. In the interior there is work by William Morris, Burne-Jones, Onslow Ford, Hamo Thornycroft and others. The special features of the church are seen in the skilful blending of red and yellow brick with stone, and in its style, which is a mingling of English Perpendicular with the Flamboyant of France. The dimensions of the church are unusually

The Roman Catholic Cathedral, Westminster

large, for it has a length of 150 feet, and a height of 60 feet.

The most original architectural effort of which London can boast is the fine Westminster Cathedral. This vast building is of red brick and in the Byzantine style. The architect, J. F. Bentley, who died before the completion of the work, reared a building of grand proportions, and one that is impressive by its simplicity. The outstanding feature is the lofty campanile tower, 283 feet in height, and the highest in the world.

Among the churches of London, the Chapels Royal occupy a unique position. They are directly connected with the Court, and are governed by the Dean and the Sub-Dean. The chapels within St James's Palace, Buckingham Palace, and Marlborough House constitute the Chapels Royal in London. The Chapel Royal, Savoy, is a Royal Peculiar, and beyond the jurisdiction of the Dean of the Chapels Royal. The chapel was built under the will of Henry VIII, and was used as a parish church down to 1717. George III made it a Chapel Royal, and it is now in the patronage of the King as Duke of Lancaster.

The Inns of Court have their own chapels, and besides the Temple Church, which is mentioned in the volume on East London, there are Lincoln's Inn Chapel, and Gray's Inn Chapel. The former was erected on part of the site of the Monastery of the Black Friars by Inigo Jones in 1623, and the latter was built in the reign of Henry VII.

18. Architecture — (*b*) Ecclesiastical. Westminster Abbey.

Westminster Abbey, or, to give its full title, the Collegiate Church of St Peter, Westminster, is the most famous building in London, and has always occupied an exceptional position among the churches of England. It is famous not only for what it is, but for what it contains. It is the national Valhalla :—

> "A place of tombs,
> Where lie the mighty bones of ancient men."

It has, moreover, the charm of architectural beauty, and the abiding associations of a thousand years. The Abbey, indeed, occupies a site that has been in use for nearly two thousand years, for, when a grave was being dug some few years ago, a Roman wall was found *in situ*; and a Roman sarcophagus found to the north of the nave may be seen in the vestibule of the Chapter House.

Westminster Abbey was a Benedictine foundation, and its abbot was one of the greatest men in the realm. As a mitred abbot he sat in the House of Lords, and gave precedence only to the abbot of St Albans. Westminster was one of the most important of the Benedictine houses, but when first the abbey church arose is uncertain. Its position on Thorney, or the Isle of Thorns, was admirably adapted for an abbey, for the soil was of gravel and sand, and the streams on either side were used as a harbour for the boats and for drainage purposes. The Thames was on the east, and so served as a means of transport for the timber and stone, besides providing the fish, for the river was then full of salmon.

The legends connected with the history of the Abbey carry us back to 616, when the first church is said to have been founded by King Sebert, whose reputed tomb is still shown in the Abbey. The church was dedicated to St Peter, and Matthew Arnold has written a charming poem on the events connected with its consecration by Mellitus, Bishop of London. It is a matter of history that this first church at Westminster was built on the site of a Roman temple to Apollo; and it is now generally accepted that it was called Westminster, because it was the minster west of the Abbey of St Mary, or East Minster in the City of London.

With regard to the early history of Westminster Abbey we have the record of a monastery being here in the tenth century; and it is quite certain that there was an important church on the site when Edward the Confessor raised his building, parts of which remain to this day. Edward the Confessor began his church in 1055, and part of it was consecrated 28th December, 1065, a few days before the Confessor passed away. The building of the nave began about 1100, and was probably finished by 1163. The new church, like the old one, was dedicated to St Peter, who was one of the Confessor's patron saints. Edward himself was canonised in 1163, when his bones were translated to a shrine. Hence it is that we connect Westminster Abbey with the Confessor, although the present building is almost entirely of a later period.

For many years the Confessor's church was large enough for all purposes, but early in the thirteenth century there began a movement in church building which is one of the most remarkable features of a wonderful century. Henry III was a lover and patron of art, but his chief passion was architecture. He was always building, and for the Abbey at Westminster it is said "he impoverished himself and London and the whole kingdom to such an extent as to bring himself into conflict with London and the nation at large."

Henry III began the rebuilding of the Abbey in 1245, and the work was still in progress up to the day of that king's death in 1272. It has been calculated that Henry spent no less a sum than £750,000 of our money on this great work. In all parts of the building everything was of the best, and it was the ardent desire of the king to surpass the finest work of England and France. He poured out his money without stint, and left the church practically as we see it to-day from the eastern apse above St Edward's shrine to the western doorway of the nave.

On the death of Henry III in 1272, the great work of rebuilding the Abbey was stopped. The Edwards cared little for the Abbey, and the first real start was made by Richard II, who was devoted to it. Henry VII, however, had most to do with the completion of the Abbey, and the last of his great works was the rebuilding of the Lady Chapel, which commenced in 1502. In 1509 Henry VII died, and this magnificent Perpendicular chapel has since been known by his name.

Henry VII's chapel shows us Gothic architecture at its best, and is far in advance of anything of the same date in England, or on the Continent. A competent modern authority says that "the vault, to begin with, is the most wonderful work of masonry ever put together by the hand of man," and Leland styled this Royal Chapel an "Orbis miraculum."

The later history of Westminster Abbey may be briefly told. In 1539 the convent was dissolved, and the treasures of the church were carried off by Henry VIII. The abbot and monks were replaced by a dean and twelve prebendaries, and in 1540 the church became the seat of a bishop, and for ten years was a cathedral. In Mary's reign the old religion was restored, and the Abbey was occupied by an abbot and fifteen monks; but when Elizabeth came to the throne, the abbot and monks were once again superseded by dean and prebendaries.

It will thus be seen that, having been built by many kings and in many centuries, the Abbey is in several styles, ranging from the Norman in the oldest parts, through the Early English and Decorated, to the Perpendicular of the Tudor period. Finally we have the western towers, erected from the designs of Sir Christopher Wren, in a style of mixed Grecian and Gothic.

The interior of the Abbey is in the form of a Latin cross, and is generally entered by the great triple entrance called "Solomon's Porch." Passing through this approach we stand in the north transept, and opposite, in the south transept, is the great rose-window overhead. A reference to the plan of the Abbey will show that between the

Westminster Abbey

transepts is the crossing, having the sanctuary on the east, and the nave on the west of it. Behind the sanctuary is the Chapel of Edward the Confessor, which is encircled by the ambulatory. Round the ambulatory are various chapels, a splendid series, of which the most magnificent is Henry VII's Chapel.

We will now glance at the chief features of the Abbey, following, as a rule, the order we have indicated. The north transept may be called the Statesmen's transept, and the first monument to arrest attention is that to the elder Pitt, Lord Chatham. Here, too, are the inscribed stones covering the graves of the rival statesmen, Pitt and Fox, to which Scott refers in the lines :—

> "The mighty chiefs sleep side by side;
> Drop upon Fox's grave the tear,
> 'Twill trickle to his rival's bier."

Palmerston and Canning, Castlereagh and Peel, Beaconsfield and Gladstone are among our statesmen who are worthily commemorated by statues in the north transept.

The south transept is crowded with monuments, mainly memorials of poets. It goes by the name of Poets' Corner. Many of the monuments are of people who were not buried in the Abbey, nor were in any way connected with its history. Here was buried the poet Chaucer, who had lived in the precincts as clerk of the works to the king. He died poor in 1400, and his ashes were transferred to the existing tomb in 1500. There are monuments to Spenser and Milton, to Shakespeare and Ben Jonson, to Dryden and Burns, and to

Shrine of Edward the Confessor, Westminster Abbey

many another poet of more or less merit. On the floor we notice that a red slab marks the grave of Browning, and a black slab that of Tennyson. Truly a goodly company of sweet singers is commemorated in Poets' Corner. The inscriptions are, however, of varying merit. On Ben Jonson's slab is the terse inscription, "O rare Ben Jonson!" and on Gay's tablet is the flippant couplet:—

> "Life is a jest, and all things show it,
> I thought so once, and now I know it."

The railed sanctuary raised on steps is beneath the central tower or lantern. The altar and reredos are modern, and the pavement is of coloured marbles. Here is the oldest contemporary portrait of any English sovereign. It is of Richard II and is probably a good portrait of that monarch, although it has been much restored. The three finest monuments in the Abbey are in the sanctuary, and are to Edmund Crouchback, Aveline, Countess of Lancaster, and Aymer de Valence.

St Edward's Chapel, or Chapel of the Kings, has the shrine of St Edward in the centre, and this is encircled by the tombs of the Plantagenet kings and Henry V. In this chapel there is the coronation chair made for Edward I to hold the stone of Scone, and the sword and shield of state said to have been carried before Edward III in his French wars. This chapel may be considered in many respects as the most interesting of all the chapels, and one recalls the lines of Francis Beaumont:—

Coronation Chair, Westminster Abbey

" Mortality, behold and feare,
    What a change of flesh is here!
    Think how many royal bones
    Sleep within this heap of stones;
    Here they lye, had realmes and lands,
    Who now want strength to stir their hands."

More than any other part of the Abbey this chapel
remains as Henry III its second founder left it. It was
that art-loving king who lavished wealth in making the
rich and glorious shrine for the relics of the Confessor,
and it seems only fit and proper that it should be marked
out, as it were, by its height above the rest of the Abbey.

The ambulatory need not detain us long. It has the
reputed tomb of King Sebert, traditional founder of
West Minster in the seventh century, and also the iron
grille protecting the tomb of Queen Eleanor. We have
not space to give details of the chapels around the
ambulatory, but we may quote Washington Irving, who
wrote thus of them: "I wandered among what once
were chapels, but which are now occupied by tombs and
monuments of the great. At every turn I met with
rare illustrious names, or the cognizance of some powerful
house renowned in history. As the eye darts into these
dusky chambers of death, it catches glimpses of quaint
effigies; some kneeling in niches as if in devotion; others
stretched upon the tombs, with hands piously pressed
together; warriors in armour as if reposing after battles;
prelates with crosiers and mitres; and nobles in robes and
coronets lying as it were in state."

We have already mentioned that of all the chapels in

Westminster Abbey, Henry the Seventh's is the most magnificent, and it is also one of the finest Perpendicular buildings in England.   It is approached by a flight of steps, and consists of a nave and two aisles, with little chapels round the apse.   Overhead is a wonderful fan vault, said to be "the greatest achievement in mason-craft in the whole world."   On either side of the chapel are the stalls, once occupied by the monks, and the banners above are those of the Knights of the Bath.   In the centre of the chapel are the tombs of Henry VII and his wife Elizabeth of York, and almost all the sovereigns of England from the time of the first Tudor king have been buried here.

The nave of the Abbey has been called "a veritable city of the dead," for it contains every kind of memorial—bust, statue, tablet, and tomb.   Memorial slabs mark the graves of Darwin, Sir John Herschel, and Lord Kelvin, and there are busts or statues of Wordsworth, Keble, Kingsley, and Matthew Arnold.   In the centre of the nave the greatest of African travellers is buried, and the single word "Livingstone" is his epitaph.

Leaving the interior of the Abbey we pass out into the cloisters on the south side.   The central garth has always been a lawn or a garden, and the present cloisters are on the site of the Norman structure.   There are effigies of several of the early abbots; and at the south-east corner of the cloisters are remains of Edward the Confessor's buildings, including the chapel of the pyx, which was originally the abbey treasury, and contained the "pyx," or box in which the standards of gold and silver were kept.

From the east cloister we reach the chapter house, a beautiful octagonal structure, supported by massive flying buttresses. It was built by Henry III, and all round it are seats for the monks, who used it for 300 years. It is, however, historically interesting as the first home of the House of Commons, and was so used till the end of the reign of Henry VIII. The chapter house has been carefully restored, and its windows filled with stained glass. It has some glass cases containing fragments of sculpture, coins, and ancient documents connected with the Abbey.

Here we must close our brief survey of Westminster Abbey, so well called by Macaulay "that temple of silence and reconciliation where the enmities of twenty generations lie buried."

Exactly in front of the Abbey is the church of St Margaret, which in this position serves as a foil to bring out the grand proportions of the structure behind it. Down to 1858, this church used to be attended four times in the year by the members of the House of Commons in state; and even now on particular occasions it serves as the official church of the Commons. The present church was built by Edward I on the site of an earlier church of Edward the Confessor, but was much restored and improved in the nineteenth century. Perhaps the most interesting feature in St Margaret's is the beautiful east window depicting the crucifixion. It is noteworthy that Caxton and Sir Walter Raleigh were buried here, and there are windows in memory of both these great men and other celebrities.

19. Architecture—(c) Domestic. Royal and Episcopal Palaces: The Tower, Westminster, Whitehall, The Savoy, St James's, Kensington, Buckingham, Lambeth, and Fulham. Houses: Staple Inn, Holland House, etc.

It will be remembered that, in the time of the Heptarchy, London was the capital of the kingdom of Essex, while Winchester occupied a higher position as the seat of the Wessex kingdom. After the Norman Conquest, however, London became the capital of England and the royal city, for William the Conqueror recognised its pre-eminent claims, owing to its river situation, by building the Tower, which was his palace. In course of time the Tower became merely a state prison, although it retained its palatial name and rank till the reign of Elizabeth.

The second palace of London was built at Westminster, under the shadow of the Abbey, by William Rufus. Of this Westminster palace little remains except the fine Westminster Hall, rebuilt by Richard II on the foundations of the original hall of William II. This, however, is sufficient to give us an idea of the scale of Norman taste and hospitality, and we are not surprised to find that Westminster Palace continued to be the chief residence of English kings for the greater part of the next five centuries. In the reign of Edward VI the chapel of St Stephen, belonging to this Norman palace, was given up to the House of Commons, and from this time forward to its destruction by fire in 1834 it was the seat of the Houses of Parliament, and of the Law Courts till 1882. Even now we speak of the Houses of Parliament as the Palace of Westminster.

Whitehall became the palace of the kings of England in the reign of Henry VIII and so continued to that of William III. It was originally called York House, and was handed to Henry VIII on the disgrace of Cardinal Wolsey, Archbishop of York. Henry VIII's Whitehall Palace was a building in the Tudor style of architecture, something like Hampton Court, with a succession of galleries and courts, a large hall, chapel, and banqueting-house. James I intended to rebuild the whole palace, and Inigo Jones prepared designs for a magnificent structure, but nothing came of it beyond the present banqueting-house. This is deservedly looked upon as a model of Palladian architecture, and one of the finest buildings in London. In the reign of William III, the whole of Whitehall, except the banqueting-house, was destroyed by fire. The present banqueting-house was built between 1619 and 1622 from the designs of Inigo Jones, and it may be noted as one of the ironies of history that Charles I, son of James I, "was led all along the galleries and banqueting-house, and there was a passage broken through the wall, by which the king passed unto the scaffold."

The Savoy Palace was built in 1245 by Peter of Savoy. It was burnt by Wat the Tiler and his followers in 1381, but was afterwards restored. The Savoy Conference was held here in 1661, and the building was pulled down in 1817-19.

The present London palaces of our king are St James's Palace, Buckingham Palace, and Kensington Palace. St James's Palace is an irregular brick building at the

The Savoy Palace, 1661

bottom of St James's Street, and was the only London palace of our sovereign from the time of the destruction of Whitehall in the reign of William III to the occupation of Buckingham Palace by Queen Victoria. It was altered or rebuilt by Henry VIII, who annexed the present park, enclosed it with a brick wall, and thus connected it with the palace of Whitehall. Little remains of the old palace except the brick gateway facing St James's Street, but there are many memories connected with it. Queen Mary I died here, and Charles II was born here. It was here, too, that Charles I passed his last night before his execution, walking the next morning "from St James's through the park, guarded with a regiment of foot and partisans"

St James's Palace

to the scaffold before Whitehall. Most of the Georges lived at St James's Palace, and although it is no longer a royal residence, the British Court is still officially known as the "Court of St James's."

Kensington Palace is a large and irregular building, which was bought by William III from the Earl of Nottingham. The higher storey was added by William III from the designs of Wren, who also planned the orangery, a very fine detached building and a masterpiece of architecture of the period. William III and Queen Mary, Queen Anne, and George II died here, and Queen Victoria was born here in 1819, and held her first Council in 1837. A large part of Kensington Palace is now used as a residence for members of the royal family; but the State Rooms, by the command of Queen Victoria,

Kensington Palace

were thrown open to the public on the occasion of her eightieth birthday. King Edward VII recently allowed the public to use the gardens, which form a favourite resort in summer-time for thousands of Londoners.

Buckingham Palace, in St James's Park, was begun in the reign of George IV, and is the most modern of the royal palaces. It is on the site of Buckingham

House, an old mansion of the Duke of Buckingham,
and was completed in the reign of William IV.
When Queen Victoria came to the throne several altera-
tions and additions were made to it, and the palace
became the royal residence in July, 1837. Since that
date it has been further enlarged, and is now the largest
of the royal palaces. The chief front is that which faces

Buckingham Palace

west, overlooking the beautiful gardens and grounds.
The east front was added in 1846, and is somewhat heavy
n effect. The whole building is in the form of a large
quadrangle, and though not beautiful from an architectural
point of view, is imposing from its size. The state rooms
are magnificent; the throne room, 60 feet long, is
decorated in crimson satin and gold, its ceiling being

lavishly adorned and painted. The principal events of the London season are the "Courts" and occasional state balls, which are attended by as many as 2000 persons. The state rooms and the private apartments contain many interesting royal portraits and other pictures. Of the national memorial to Queen Victoria designed by Sir Aston Webb, R.A., with its statue of Queen Victoria by Mr Thomas Brock, R.A., embellished with allegorical figures of Truth and Justice, and a group emblematic of Motherhood we have already spoken.

There are two episcopal palaces in London—Lambeth the home of the Archbishop of Canterbury, and Fulham, the residence of the Bishop of London. The earliest reference to the former is in the reign of Edward the Confessor, when the Manor of Lambeth was given to the see of Rochester, but in 1197 it was exchanged for the Manor of Dartford, which was in the possession of the Archbishop of Canterbury. Since that date Lambeth Palace has been the London residence of the Primate, and has played an important part in our history.

The present building exhibits various gradations in its architecture from Early English to late Perpendicular. It is entered through a Gothic gatehouse of red brick, the lower floor of which was used as a prison. The chapel is the oldest part of the building, and was built by Boniface, Archbishop of Canterbury, from 1244–1270. It is Early English in style, with lancet windows, crypt, and a gorgeous modern groined roof. In this chapel all the Archbishops since the time of Boniface have been consecrated, and although it suffered much

Lambeth Palace

during the Cromwellian period, it has since been carefully restored. The Lollards' Tower at the west end of the chapel dates from the middle of the fifteenth century, and is so called from the Lollards, who are said, incorrectly, to have been imprisoned in it.

The hall of the palace was built by Archbishop Juxon, who attended Charles I to the scaffold. The open oak roof is the finest feature of the hall, and the walls are hung with a long series of portraits of the Archbishops. The library was founded by Archbishop Bancroft, and is rich in historical and state manuscripts, illuminated service books and missals. Here we must leave Lambeth Palace, for it has been said, with a considerable amount of truth, that " a complete history of the Archbishop's residence at Lambeth would be a history of England."

We may now turn to Fulham Palace, which is said to be the oldest inhabited house in England. Whether this be so or not, we know that the Manor of Fulham was granted in 831 to Erkenwald, Bishop of London, and during all the intervening period, with the exception of a few years in the Cromwell *régime*, it has remained in the possession of the See of London. The palace dates in part from the reign of Henry VII, and though of no architectural pretensions, it looks impressive from its antique appearance. The building consists of two quadrangles, and the well-wooded grounds are enclosed in a moat nearly a mile in circumference. The library, once used as the chapel, is perhaps the most interesting part of the building. Its walls are hung with portraits of all the Bishops of London from the Reformation to the present

Fulham Palace

day, and in its windows are the armorial bearings of various prelates.

Now, leaving the royal and episcopal palaces, we may pass on to consider some of the more interesting houses in the western portion of London. The City of London is the centre of the business of the metropolis, and has the chief banks, exchanges, markets, and warehouses, while

Northumberland House

from the residential point of view East London is the home of the working-classes. The City of Westminster, on the other hand, is the centre of the official life of London, and, within recent years, has been largely rebuilt. Many imposing buildings have been erected, new thoroughfares laid out, and old ones widened. The broadening of the Strand is approaching completion,

and the wide thoroughfare known as Aldwych and
Kingsway has taken the place of an old and congested
neighbourhood. Where the great Northumberland House
stood, there is now an avenue of huge hotels and clubs.
The new Charing Cross Road and Shaftesbury Avenue
have opened up districts that were insanitary and over-
populated, while old-fashioned houses and offices in

Holland House

Parliament Street have given place to magnificent piles
of Government buildings. The Victoria Embankment,
the work of the latter part of the nineteenth century, is
the finest thoroughfare in the world, and has palatial
hotels, clubs, and other buildings of an ornate character
overlooking the Thames.

It will thus be seen that West London is largely a
new London, and every year there is swept away some

relic of the past. There is, however, a picturesque collection of old wooden gabled houses in Holborn that have been saved from destruction. They are known as Staple Inn, and give one a good idea of what the old streets of London must have been like in Tudor times.

The stateliest piece of Jacobean architecture within the County of London is Holland House, which stands amid large and beautiful grounds in Kensington. It is a picturesque red brick and stone building, in Renaissance style, and was built in 1607 by John Thorpe, a celebrated architect of that period. The stone gateway close to the house on the east was designed by Inigo Jones and carved by Nicholas Stone, master mason to James I. Not only is Holland House of note from an architectural point of view, but it has most interesting literary associations. Addison, Fox and Macaulay were frequent visitors at this great and famous house, which was the resort of wits and beauties, of painters and poets, of philosophers, dandies and statesmen.

The town houses of the aristocracy are in such neighbourhoods as Mayfair, Belgravia, and Park Lane, and the houses of our great nobles—Devonshire House, Stafford House, Grosvenor House, Lansdowne House, and Bridgewater House—are all of the Georgian, or even later period. From an architectural point of view they are not striking, and impressive only from their good proportions and appearance of solidity. Internally, however, many are magnificent, with fine marble staircases, beautiful decorations, and priceless works of art.

Chelsea is one of the districts that has been almost

entirely rebuilt in recent years, and here we find a rever-
sion to red brick and the old English style of architecture.
In some boroughs, such as Hampstead, there are left

Joseph Addison

houses of a really good type. They are comely, and in
many cases delightfully quaint and irregular, and may be
seen to great advantage in Church Row, which, with its

pleasant old houses of red brick and its limes flourishing in the roadway, has an old-world touch in strange contrast to the modern flats close by. Indeed one does not need to go far in London to realise not only the change of style in building, but also the change of material. Brick, and stone, and timber, are giving place to reinforced concrete and steel girders. Everything seems of a stern prosaic character, and the chisel of the mason and tool of the carver are no longer required. Modern London has no definite style of architecture. In recent years we have seen the French influence at work, to be succeeded by the Italian, which in turn is followed by some variety of Gothic. It is all a borrowing from the past or from other lands, and London has yet to form its own style.

20. Communications — Ancient and Modern. The Thames formerly the Normal Highway of London. The Thames Watermen.

For many centuries the chief highway of London was the Thames, which played a most important part in the life and history of the City. London has now developed to such an extent that the number of people who use the Thames either for business or pleasure is really very small. It is of the utmost importance, however, that we should realise that the Thames made London, for it was the most important, if not the only

highway by which merchandise in large quantities was
brought into the City from the provinces or from abroad.

The Thames then is the most natural starting-point
when one considers the communications of London. A
glance at a map of Roman London brings out clearly
one important point. A great many of the ancient roads,
both those of pre-Roman as well as of Roman days, seem
to converge towards a single point on the northern bank
of the Thames. Some of these roads, after traversing
England for hundreds of miles in almost a straight line,
are turned aside in order to reach that point. Now a little
reflection will show the reason for this diversion of route.
It was that the road might be carried over the ferry or
bridge where the Thames was the narrowest, and the
present London Bridge is nearly on the site of the first
ferry or bridge.

London owed its early prosperity to the building of
the bridge, which is the first ascertained fact in the
history of Roman London. We have already referred
to the making of London Bridge in other chapters, so that
it is not necessary to urge its importance all through
our history. We are now in a position to consider,
briefly, the roads through Roman London. The road
from the south crossed the bridge to Eastcheap, where it
divided into two branches, one of which ran northward to
Bishopsgate, and the other north-westward to Newgate.
The northward street at Bishopsgate again divided, the
westward road ran to Lincoln and York, while the eastward branch crossed the Lea at Old Ford and became
the main road through Essex. The north-westward road

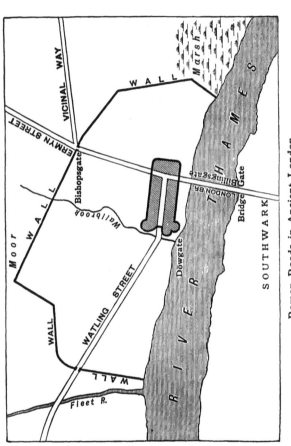

Roman Roads in Ancient London

passed from the City at Newgate, and throughout its entire length from Kent to its termination in Wales was known as Watling Street. One of the small streets in London, probably in the course of the original, still bears that name.

Besides the Thames, the Bridge, and the two or three main Roman roads, there was the Walbrook, a stream of some importance then, but now only a matter of history. Its name is given to Walbrook, a thoroughfare by the Mansion House, and when excavations are made, traces of its former channel are often found.

Now let us return for a short time to the Thames, which was for so long the normal highway of London. When the roads were few and bad, and when railways were unknown, it was considered safer and better to move from place to place by means of boats or barges. Londoners thus ran no risk of being stopped by footpads or highwaymen, and the "silent highway" of London was then used as much for pleasure as for business. It was not till the later part of the nineteenth century that the Thames was embanked, and with a little thought we are able to realise that it was once broader than it is to-day. At high tide the water came up to the busy street we now call the Strand, which was then the strand or shore of the river. When the river was not embanked, it was not easy to get into a boat when the tide was low, and thus in several places "stairs" were built which allowed persons to land or embark at all states of the tide. The popular old English song, "Wapping Old Stairs," reminds us of this fact. Water-gates, too, were erected, and these allowed the boats to come in at any time; and from the little wharves which they enclosed the owners of the gate could embark at any time in their own barges. A good example of these water-gates, as we have seen, still exists at the bottom of Buckingham Street, Strand.

As the river was so much used by all classes, it is evident that the number of boats must have been very large. Many of the citizens had their own boats, and stately barges were kept by the Lord Mayor, the City Companies, and the great nobles who then resided in their town houses along the Thames. We can form some idea of the traffic on the river in the reign of Queen Elizabeth, for Taylor, the Water Poet, tells us that in his time "the number of watermen and those that lived and were maintained by them, betwixt the bridge of Windsor and Gravesend, could not be fewer than 40,000." Later we find that the watermen were made into a Company, and that they could furnish 20,000 men for the fleet. The watermen had the sole right to carry passengers for hire upon the Thames, and were very zealous in protecting their rights. In 1850, a writer laments that "the introduction of steamboats has changed the whole character of the Company, and for every fifty watermen in the reign of Elizabeth, there is not more than one now." We may go further and remark that few steamboats now ply on the Thames, and passenger traffic has almost ceased.

The Thames is of the greatest interest in our history, and whole chapters might be written of the processions, happy and unhappy, that have passed along its stream.

How many state prisoners have passed from their trials at
Westminster to their doom at the Tower ! Some of our
greatest men have stopped in their boat outside the
Tower, and, entering its gloomy portals by way of the
Traitors' Gate, have gone to their cells, only to be beheaded after a short time on Tower Hill. We have
seen in Chapter 6 how the Thames carried the Seven

The Seven Bishops on their way to the Tower

Bishops to the Tower, and how it became the repository
for a time of the Great Seal of England, which James II
in his flight threw into the water.

When Queen Elizabeth died at Richmond, her body
was carried with great pomp by water to Whitehall, and
a contemporary poet thus writes :—

> " The Queen was brought by water to Whitehall ;
> At every stroke the oars did tears let fall " :

Cowley, the poet, died at Chertsey and his body was borne by water to Whitehall, and Pope thus commemorates this event :—

> " Oh, early lost! what tears the river shed
>   When the sad pomp along his banks were led."

Nor must we forget that a greater than Cowley was brought in state by water from Greenwich, for thus was Nelson carried to his last resting-place in 1805.

We must close our historical references to the Thames by a brief glance at Pepys. We cannot read much of the *Diary* without coming across such a phrase as, " By water to Woolwich," or " By water to Whitehall." The Thames plays a most important part in the London of Pepys, and right well did the diarist know how to amuse himself by using the river in going from one place of entertainment to another. Here are a few extracts from his *Diary* on August 23, 1662: "I walked all along Thames-street but could not get a boat ; I offered eight shillings for a boat to attend me this afternoon, and they would not, it being the day of the Queene's coming to town from Hampton Court. So we fairly walked it to Whitehall, . . . and up to the top of the new Banqueting House there, over the Thames, which was a most pleasant place as any I could have got ; and all the show consisted chiefly in the number of boats and barges ; and two pageants, one of a King, and another of a Queene. . . . Anon come the King and Queene in a barge under a canopy with 1000 barges and boats I know, for we could see no water for them, nor discern the King nor

Queene. And so they landed at White Hall Bridge, and the great guns on the other side went off."

But now it is time to leave the Thames with its merry-making, its pathos, and its tragedies, and pass to the more prosaic study of the present-day communications

St Pancras Station
(*With King's Cross Station in the distance*)

of London. The streets in the County of London are maintained for the most part by the metropolitan borough councils and the City Corporation. The London County Council maintains the roadway of the County bridges, of the Thames tunnels, and the Victoria Embankment. In the whole of the County of London there are 2135 miles of public roads and streets, and these are kept in a very high state of repair, while their cleanliness and good lighting are generally recognised.

With regard to railway communications we find that no fewer than ten trunk lines have their termini in London, while the local lines, which are now generally electrified, are numerous. The first Electric Railway in London was that from the Bank to Stockwell, which was opened in 1890, and the first of the "Tube" Railways was the Central London from the Bank to Shepherd's Bush. The most frequented underground line is the Inner Circle, which is very convenient for travelling east to west, from Aldgate to Kensington. The best way to study the railways of London is to get a good railway map of the metropolis and trace the various lines. The total length of all the lines in London is about 250 miles. There are 329 stations in the County, and about 7800 trains enter and leave London every day.

Tramways have not yet been allowed to penetrate into the heart of London, but they are largely used on the northern and southern portions. Most of the tramway lines belong to the London County Council, and with few exceptions they are electrified. The chief starting-points of the trams for South London are from the Embankment; for North London, from Moorgate Street and the bottom of Gray's Inn Road; for East London, from Bloomsbury and Aldgate; and for West London, from Hammersmith and Harrow Road. There are about 124 miles of tramways in London, and with the exception of a low-level underground tramway from Theobald's Road to Aldwych, they are all above the surface.

The traffic of London is at present in a state of transition owing to the advent of motor omnibuses. These

Blackfriars Bridge
(*Showing tramway*)

have almost superseded the horse-drawn vehicles, and although they were not introduced till 1899, they are nearly 2000 in number. The omnibus has long played a most important part in London locomotion, and from the top of a 'bus one gets some insight into the life of the City.

We now come to the last means of locomotion in London. The London cabs have taken the place of the old Hackney coaches, and are now of three kinds—the four-wheeler, the two-wheeler, or "hansom," and the motor "taxi-cab." The latter class is rapidly increasing, while the hansom will soon be a thing of the past.

No city in the world is so well provided with the means of locomotion as London, and its history allows us to compare the state of affairs at various periods. London was famed for its coaching-houses till the advent of railways, and the coaches that ran from London to all parts of England were noted for their speed. In the reign of Charles II the fast coaches were called "Flying Machines," and here is a contemporary advertisement concerning them : "All those desirous to pass from London to Bath, or any other Place on their Road, let them repair to the Bell Savage on Ludgate Hill in London, where they may be received in a Stage Coach which performs the whole journey in Three Days (if God permit) and sets forth at five in the Morning. Passengers to pay One Pound five shillings each, who are allowed to carry fourteen Pounds Weight—for all above to pay three half-pence per Pound."

Things are different now and we can leave London and reach Bath in less than two hours. The journey of 107 miles only costs 8s. 11d., and we have no need to dread attacks on the highway by robbers or footpads.

21.   Administration   and   Divisions —
The City of Westminster.   The
London County Council. The Port
Authority.   Trinity House.

London is the youngest of all our counties, and had
no central representative government till 1889, when the
first London County Council was constituted.   From
1855 to 1888 the Metropolitan Board of Works was the
chief authority, but as it was not a popular body its work
was not altogether satisfactory.   In this chapter we have
to consider the constitution and work of the London
County Council, which is the chief authority for the
Administrative County of London.   The constitution and
work of the Corporation of the City of London, which has
jurisdiction over a very small but most important area, are
considered in the volume on East London.

The London County Council consists of 118 elected
representatives and 19 aldermen, and the election takes
place in March, every three years.   The aldermen are
elected by the Council for six years, so that nine retire
at the end of one period and 10 at the next period.
The first Chairman of the London County Council was
Lord Rosebery, and other men of eminence have since
been elected to this high position.

The work of the London County Council is of a
most important character.   One of the most interesting
features of the Council is the care and development of
the parks and open spaces, and of them some account is

Design for the London County Hall

given in another chapter. The Council maintains the London Fire Brigade, which is one of the most efficient in the world; and it has important duties in connection with the health of the people, for it controls slaughter-houses, dairies, cow-sheds, and lodging-houses. It is constantly making improvements, by clearing insanitary areas, and by widening streets. It manages the lunatic asylums of London, and has a staff of officers to supervise weights and measures. It is the chief authority for education, and superseded the London School Board in 1903. Among the various other duties of the London County Council may be mentioned those relating to the management of the tramways, the housing of the working classes, and the preservation of historic buildings.

Besides the two chief governing bodies of London, the Administrative County has 28 Borough Councils, which by the Act of 1899 superseded the vestries and district boards. The council of each borough consists of a Mayor, and not more than 10 aldermen and 60 coun-cillors. The powers and duties of these borough councils are not so important as those of the London County Council; but they are concerned with the maintenance of roads and their cleansing and lighting, with public libraries, baths, and wash-houses, and with other useful work within their area.

The County of London has 31 Boards of Guardians, four Boards of Managers of School Districts, and two Boards of Managers of Sick Asylum Districts. These various bodies are mainly concerned with Poor-law ad-ministration, that is, with the management of workhouses, and with the work of looking after the poor, sick, and aged.

For the administration of justice, London has a Court of Quarter Sessions, and 15 Courts of Petty Sessions. There are also 14 Police Courts with Magistrates, and 14 County Courts. The Central Criminal Court has jurisdiction not only over all London and Middlesex, but also over parts of Essex, Kent, Surrey, and Hertford. There are prisons at Brixton, Holloway, Pentonville, Wandsworth, and Wormwood Scrubs, and the police force in London numbers about 18,000 men.

The County of London is in the dioceses of London, Southwark, St Albans, and Canterbury. The Archbishop of Canterbury has his official residence at Lambeth Palace, and the Bishop of London has Fulham Palace, and London House, St James's Square. Formerly the ecclesiastical parish coincided with the civil parish, but now, while there are 69 civil parishes, there are 610 ecclesiastical parishes in London.

For parliamentary purposes London is divided into 58 constituencies, with one member for each, except in the case of the City, which returns two members.

The Metropolitan Water Board was formed in 1902, and has control of the water-supply, but its jurisdiction extends far beyond the County of London. The Port of London Authority was established in 1908, for the purpose of administering, preserving, and improving the Port of London, and superseded to a large extent the Thames Conservancy Board, which is now concerned mainly with the upper part of the Thames.

London is not only first among British seaports but is the greatest seaport in the world. For some years past, however, there has been a feeling that London as a port has not been making the same rate of progress as such continental ports as Hamburg, Antwerp, Bremen, Rotterdam, and Havre. In 1908 a Bill was passed " to provide for the improvement and better administration of the Port of London and for purposes incidental thereto." As a result a Port Authority has been established, which consists of 18 elected and 12 appointed members. This body has taken over the docks and has power to construct, maintain, and manage any docks, quays, wharves, jetties, and piers that may be necessary. The Port Authority has now jurisdiction over the Port of London, which extends from Teddington Lock to an imaginary straight line drawn from Havengore Creek in Essex to the Land's End at Warden Point in Sheppey, Kent.

Besides the new Port Authority, there are two other bodies which have jurisdiction in particular cases over the same area. The City Corporation is the Port Sanitary Authority from Teddington Lock seawards; and Trinity House has charge of the pilotage, lighting, and buoying of the river from London Bridge seawards. There are about 350 licensed pilots in London and the fees received for pilotage amount to nearly £150,000 per annum.

22. Public Buildings — (*a*) Parliamentary and Legal. The Houses of Parliament, Royal Courts of Justice, Inns of Court.

The Houses of Parliament are on the left bank of the Thames between the river and the Abbey. They

The Houses of Parliament

form one of the most magnificent buildings ever erected in Europe, and the largest Gothic edifice in the world. The buildings occupy the site of the old Royal Palace at Westminster, which was burnt down on October 16, 1834. They cover an area of nearly eight acres, and their cost has exceeded £3,000,000. The architect was

Sir Charles Barry, and the first stone was laid on April 27, 1840. The building, however, was not completed till 1860, although the first session of Parliament in the new Palace was in 1852.

Although Barry liked the Classic rather than the Gothic, he chose the Perpendicular as the most suitable for this purpose, and he wisely decided to incorporate Westminster Hall, which was a remnant of the old Palace, into the new buildings. In some of the details of the exterior work we are reminded of the beautiful Town Halls of Brussels and Ghent. The stone employed for the external masonry is a magnesian limestone, while the river terrace is of Aberdeen granite. There is very little wood about the building, all the main beams and joists being of iron. The river front may be considered the principal. This magnificent facade, 900 feet long, is divided into five compartments, panelled with tracery, and decorated with rows of statues and coats of arms of English monarchs from the Conquest to the present time.

The chief features of the building are the three towers which break the skyline of the Palace. The greatest of these is the Victoria Tower, said to be the loftiest in the world, rising as it does to a height of nearly 340 feet. This tower contains the Royal Entrance, of which the roof is a rich and beautifully-worked groined stone vault, while the interior is decorated with the statues of patron saints of England, Scotland, and Ireland. From the flagstaff on this tower the Union Jack is hoisted whenever Parliament is actually sitting.

The Clock Tower is a few feet lower than the Victoria Tower, but in the eyes of Londoners it probably takes first rank. It contains " Big Ben," which weighs about 14 tons, and the Palace Clock which strikes the hours and chimes the quarters upon eight bells, while it shows the time upon four dials each having a diameter of 30 feet. The Clock Tower has been used as a kind

House of Lords

of prison, for within one of its chambers, members who have incurred the displeasure of Parliament have been confined. From the summit of the Clock Tower the light of a powerful lantern during the dark hours notifies that Parliament is sitting.

The Central Tower, the third of the group, is graceful, and although it formed no part of the original design,

it was subsequently added by Barry in order to carry out
a scheme of ventilation.

Having glanced at the exterior of the Houses of
Parliament, we may enter this fine building either
through Westminster Hall or Old Palace Yard, each of
which leads into the Central Octagon Hall, from which
the right-hand passage leads to the House of Lords, and
the left to the Commons.

The Mace and Purse, House of Lords

The House of Peers, the "Gilded Chamber," is a
noble room of unusual magnificence. No expense was
spared to make it one of the richest chambers in the world.
The spectator, however, is hardly aware of the lavish
richness of its fittings from the masterly way in which all
are harmoniously blended. The architect intended to
make it grand, for it was not only to be the meeting

place of the Peers, but the Audience Chamber of the Sovereign. At the south end are the thrones for the King and the Queen, and two state chairs for the Prince and Princess of Wales. In front of the Throne is the Woolsack, on which the Lord Chancellor sits. The walls have some fine historical frescoes; the twelve windows are gorgeous with portraits of our monarchs, and the

House of Commons

niches between them have statues of the 18 barons of Magna Carta fame.

The House of Commons is not so ornate or dazzling as the House of Peers, but it has fine oak panelling and stained glass windows. At the north end is the chair for the Speaker, who controls the debates when the House is not in Committee. There are galleries for visitors, for

the public, and for the reporters, but the floor of the
House has not accommodation for the whole of the 670
members.   When there is a full House, many of the
members overflow into the galleries.   In front of the
Speaker's Chair is a table for the clerks, and at the south
end of this table, when the Speaker is in the Chair, rests
the Mace, the symbol of the Speaker's authority.   The
division lobbies are on the east and west of the
Chamber, and into these members pass when a division
is taken and record their votes—the Ayes going to the
Speaker's right and the Noes to the left.

The Central or Octagon Hall is midway between
the House of Commons and the House of Peers.   It is
a grand apartment, with a stately vaulted stone roof
containing more than 250 carved bosses, and it contains
statues of great Parliamentarians of recent years.   Leading
from the Central Hall to Westminster Hall is St Stephen's
Hall, where are statues of "men who rose to eminence
by the eloquence and abilities they displayed in the House
of Commons."   It derives its name from occupying the
same space as St Stephen's Chapel of the ancient Palace,
which was the old House of Commons.   Beneath
St Stephen's Hall is the old crypt of St Stephen's
Chapel, and this has been recently restored as a place of
worship for the residents of the Houses of Parliament.

The Houses of Parliament with their eleven courts,
one hundred staircases, and eleven thousand rooms have
been criticised rather severely.   Mr Ruskin quoted this
building as a superb instance of the want of imagination
shown by English architects in raising a building entirely

decorated by "straight lines." On the other hand, the majority of English people look upon the new Palace at Westminster as one of the most pleasing buildings in London, and the French critic, M. Taine, supports this popular view. He says :—"The architecture has the merit of being neither Grecian nor Southern ; it is Gothic, accommodated to the climate, to the requirements of the eye. The palace magnificently mirrors itself in the shining river ; in the distance its clock tower, its legions of turrets and of carvings are vaguely outlined in the mist. Its soaring and twisting lines, complicated mouldings, trefoils, and rose windows diversify the enormous mass, which covers four acres, and produces on the mind the idea of a tangled forest."

The Royal Courts of Justice, more commonly known as the Law Courts, are a vast and magnificent pile of buildings in the Gothic style, on the north side of the Strand, just beyond Temple Bar. Some idea of the extent of them may be gained when we know that there are no less than 19 Courts, and about 1100 apartments of all kinds. At the present time the accommodation is insufficient, and an enlargement of these fine buildings is taking place.

The architect of the Law Courts was George Edmund Street. He did not live to see the completion of his work, but died, worn out by his strenuous labours, in 1881, and the completion of the building was accomplished by his son, Mr Arthur Edmund Street, and Sir Arthur Blomfield. The building operations actually began in 1874, and it was not till December 4, 1882, that Queen Victoria, with great ceremony, formally opened the Royal Courts of Justice.

It is generally admitted that the Law Courts are Street's greatest achievement, but the splendid front to the Strand of 500 feet suffers from its proximity to the crowded thoroughfare. A competent authority says

The Law Courts

that this front " has no lack of effectiveness, and its towers and turrets, its arcades, its oriels and gables, its polished pillars and pilasters, lend to it a variety not inconsistent with unity." The chief feature of the structure is the Great Hall, a noble specimen of Early English Architecture, with a finely groined stone roof, a mosaic floor, and a deeply recessed and elaborately decorated entrance, above which is a lofty gable containing a rose window of rare beauty.

It is of historical interest to remember that the Courts of Law were formerly held at Westminster Hall and Lincoln's Inn. Thus we find as early as the reign of Edward I that the Courts of King's Bench, Common Bench, and Exchequer were all sitting in Westminster Hall, and from the reign of Edward II the Court of Chancery was held in Lincoln's Inn. If we go back to the time of the Norman kings, we find that the Exchequer and the " Curia Regis," two of the royal courts, followed the king from place to place, and it was not till the time of Magna Carta that it was decided that the Court of Common Pleas should be fixed at Westminster.

Westminster Hall, the old Hall of the Palace of our kings at Westminster, was incorporated into the present Houses of Parliament. It was originally built by William Rufus, but the present Hall was repaired, so as to be almost a new structure, in the closing years of Richard II. The roof is of oak, and the first of its kind in the country. Some of our early Parliaments were held in this Hall, and it is curious that the first meeting of Parliament in the new building was to depose the very king by whom it had been rebuilt. The Law Courts of England were permanently established in Westminster Hall in 1224, and from that date onwards to the opening of the new Law Courts in the Strand, they were held either in that Hall or in certain courts adjoining it. This venerable Hall has been the scene of many great trials. Here

Wallace was tried and condemned, and here, too, Sir Thomas More was doomed to the scaffold. Here the great Earl of Strafford went through his trial, and here, also, his king, Charles I, was arraigned before the High Court of Justice. Here it was that the Seven Bishops were acquitted, and here Warren Hastings was tried and also acquitted, notwithstanding the impassioned eloquence

Westminster Hall

of Burke and Sheridan. Westminster Hall is, indeed, the most historic of all our buildings connected with the administration of justice.

We may now pass to a brief review of the Inns of Court which have been styled "the noblest nurseries of Humanity and Liberty in the Kingdom." They are four in number—Inner Temple, Middle Temple, Lincoln's Inn,

and Gray's Inn—and they are situated in the neighbourhood of Fleet Street, Chancery Lane, and Holborn. These
ancient buildings, with their courtyards and grass-plots,
give a charm to the heart of London and remind us of the

Thomas Wentworth, Earl of Strafford

colleges of Oxford and Cambridge. They were named
Inns from the ancient custom of masters receiving law
apprentices to board and reside with them ; and they
were called Inns of *Court* from being formerly held in
the " Aula Regia," or Court of the King's Palace.

Lawyers are still the principal dwellers in these Inns of Court, and lectures and examinations take place in them. Their government is vested in "Benchers" consisting of the most successful and distinguished members of the English Bar, and "keeping commons" by dining in Hall

Lincoln's Inn Gateway

is still an indispensable qualification for being called to the Bar.

In the volume on East London reference is made to the Inner Temple and the Middle Temple, so that we need only consider Lincoln's Inn and Gray's Inn in this chapter. Lincoln's Inn is one of the most important Inns of Court and is called after Henry de Lacy,

Earl of Lincoln, whose town house occupied a portion of this site. The Gateway of brick in Chancery Lane is the oldest part of the present building, and bears on it the date of 1518. It is said by Fuller that Ben Jonson, a poor bricklayer, was found working on this structure with a trowel in his hand and a Horace in his pocket. Lincoln's Inn Chapel, in the Perpendicular style, was built by Inigo Jones in 1623. There is some good stained glass in the windows, while the crypt beneath the chapel, on open arches, was built as a place for the students and lawyers "to walk in and talk and confer their learnings." Lincoln's Inn Hall is a noble structure in the Tudor style. It is of red brick with stone dressings, and has a fine roof of carved oak. The Library is the oldest in London, and has a priceless collection of manuscripts and books. Among the distinguished members of Lincoln's Inn may be mentioned Sir Thomas More, Shaftesbury, Oliver Cromwell, and William Pitt.

Gray's Inn, the most northerly of the Inns of Court, is called after Lord Grey de Wilton, of the time of Henry VII. There is a great charm in Gray's Inn with its quiet old garden and its fine spreading trees. The Hall was built in 1560, and the gardens were first laid out about 1600. Bacon, who dates the dedication of his *Essays* "from my chamber at Graies Inn," and Lord Burghley, are two of the great worthies of the Inn. Bradshaw, who presided at the trial of Charles I, was a bencher of this Inn. Gray's Inn Gardens, in the days of the later Stuarts, were a fashionable promenade on a summer evening. Here it was that Pepys the diarist and his wife used to walk on

In Gray's Inn Gardens

Sundays " to observe the fashions of the ladies." Bacon is said to have planted some of the trees, but probably none now remain coeval with his time.

Besides the four Inns of Court there were formerly nine Inns of Chancery attached to them; but of these some have entirely disappeared, and of others only vestiges are left. Clement's Inn has gone, and the special feature of its garden, a sundial upheld by the kneeling figure of a blackamoor, is now in the Temple Gardens. Barnard's Inn has been converted into a school by the Mercers' Company. Staple Inn is familiar from the timbered, gabled front it presents to Holborn. Furnival's, Thavies', and the other Inns famous in olden days are no more, and their quiet little gardens have shared their fate.

23. Public Buildings—(b) Government and Administrative Offices in Whitehall and Parliament Street. Somerset House. Spring Gardens.

Whitehall, with its continuation Parliament Street, was once a street of palaces; now it is almost entirely given over to palatial Government Offices. By their erection this part of London has been much improved, and in the opinion of competent judges, this thoroughfare from Charing Cross to Westminster is the finest in London.

As the most important of the Government offices are on the west side of Whitehall, we may well begin with the Admiralty, which is the most northerly. This building dates from the early eighteenth century, but owing to the rapid expansion of the navy in the later part of the nineteenth century new buildings became necessary. These were completed in 1905, and form a quadrangle behind the old building, one of the blocks overlooking the Horse Guards Parade. The old Admiralty's most famous

The Horse Guards

association is with Nelson, for here the body of the great sailor lay in state during the night of January 8, 1806. The new Admiralty buildings have wireless telegraphy apparatus installed on the towers, and by this means communication is ensured with ships within a radius of 1600 miles.

The Horse Guards is the next building and dates

from 1753. The Commander-in-Chief had his offices here, before the post was abolished. The passage under the clock tower leads to the Horse Guards Parade, where the picturesque ceremony of the Trooping of the Colours takes place on the King's birthday.

Beyond Dover House, which has the offices of the Secretary for Scotland, is a large block of Government offices containing the Treasury and the Privy Council Office. It is of interest to note that the Treasury has been situated in this locality since the time of Charles II, and that the present building was erected by Barry in 1846.

Further to the south, where Whitehall merges into Parliament Street, is Downing Street, named after Sir George Downing, soldier and politician, of the time of Charles II. "Downing Street" is now synonymous with the seat of government in the British Empire, and "No. 10" has exceptional interest as the temporary residence of successive Prime Ministers. Walpole was the first Premier to occupy it, and here, among others, have resided William Pitt, Gladstone, and Beaconsfield. The official residence of the Chancellor of the Exchequer is next door, and this house was for many years the home of Gladstone.

South of Downing Street is another great block of buildings comprising the Home Office, the Foreign Office, the Colonial Office, and the India Office. These are ranged around a quadrangle and were designed by Scott in the Italian style. The Foreign Office is the most splendid part of this block, and its grand staircase is the chief feature of this building. There is a magnificent conference room, and sometimes Cabinet Councils are held in the Foreign Office.

The lower part of Parliament Street on the west was demolished in 1899 to make way for new Government offices, which are occupied by the Local Government

The Foreign Office

Board and the Board of Education. These buildings form the finest that have been erected in recent years, and are in the style of the Italian Renaissance. A further extension is now in progress, and the new buildings will be assigned to the Board of Trade.

We will now pursue our course and note the Government buildings on the east side of Parliament Street and

Whitehall. The first of importance are the offices of New Scotland Yard, which have been the headquarters of the Metropolitan Police since 1891. Passing the Banqueting Hall, which is noticed elsewhere, we reach the new War Office, a vast block of buildings completed in 1906. It is built of Portland stone, in the Renaissance

The War Office

style, with domed towers at each corner. In front of the War Office is an excellent equestrian statue of the Duke of Cambridge, who was Commander-in-Chief of the British Army from 1856 to 1895. Beyond the War Office, in Whitehall, is the newest of the Government offices, which is used by the Department of Woods and

Forests. We are now near Trafalgar Square, and looking down the street we have a fine view of the Government offices on either side, and the grey mass of Westminster Abbey in the distance.

Leaving the Government offices in Whitehall and Parliament Street, we may pass to Somerset House, which is entered from the Strand. The building is in the form

Trafalgar Square looking N.W.

of a quadrangle, and the river terrace, 800 feet long, built after the Venetian style, is its chief feature. The present Somerset House was erected in 1776 on the site of a palace of the Protector Somerset, built in 1549. The eastern wing of Somerset House is now King's College; while the other portions are the offices of the Inland Revenue, the Registrar-General of Births, Deaths, and

Marriages, and the Will Office. All the wills of the kingdom are kept here, and the wills of many noted men may be seen, notably those of Shakespeare, Newton, and Dr Johnson. The registers of wills go back to the fourteenth century and visitors are allowed to examine them under certain regulations.

The present headquarters of the London County

Somerset House

Council are known as Spring Gardens, or more properly Spring Garden. This is situated between Charing Cross and the Mall, and is named after some springs in the neighbourhood. The London County Council took over these offices from the Metropolitan Board of Works, and enlarged them for their new purposes. The accommodation, however, is quite inadequate for the greatest municipal authority in the world, and a new and palatial County Hall (vide p. 175) is being erected at a cost of nearly £2,000,000 on the Lambeth side of the Thames.

24. Public Buildings—(c) Museums and Exhibitions.   British   Museum, Natural History Museum, Victoria and Albert Museum, India Museum, Imperial Institute.

The British Museum in Great Russell Street, Bloomsbury, and the Natural History Museum at South Kensington, are both under the same management.   It is well to remember that, before 1883, the nucleus of the fine collections now at South Kensington was formed at the British Museum, and these were then removed to give more space for the other exhibits that had accumulated.

We will begin with the British Museum, which Ruskin said is "the best ordered and pleasantest institution in all London, and the grandest concentration of the means of human knowledge in the world."   It will be gathered from this summary how impossible it is in our space to do little more than give a short sketch of the origin and growth of the museum, and to indicate some of its principal departments and their contents.

The present building is on the site of Montague House, which was acquired in 1759 to house various collections which had been left to the nation.   This old house proving quite insufficient for its purpose, the main portion of the museum, designed by Sir Robert Smirke, R.A., was completed in 1845. The great reading-room with its vast dome was added in 1857, and large additions have recently been made to the building, King Edward VII having laid the memorial stone of a new wing on the north side. The British Museum is of Ionic

The British Museum

architecture, and is considered one of the most successful imitations of the Greek in our country. It is faced with a portico, whose columns extend round the wings of the building ; and the sculpture in the pediment is by Sir Richard Westmacott. The dome of the reading room is slightly larger than that of St Peter's at Rome, and much larger than that of our own St Paul's, but as it is less lofty than either of those structures, and is so hemmed in by other buildings, it is almost impossible to get a good view of it from the outside.

On entering the museum we notice in the hall some pieces of modern and classical sculpture, besides Greek and Latin inscriptions.   We pass on to the large reading room, which contains on its shelves 80,000 books, and has seating accommodation for hundreds of readers, who are pursuing researches into all kinds of literature.   The British Museum library consists of more than 2,000,000 volumes, and this number is being annually increased, as a copy of every book published in the United Kingdom must be sent to the museum.   Students and readers have practically the right of access to all these books, which occupy over 40 miles of shelves.   This library, which is the second largest in the world, has books which are not only rare but unique, and the specimens of beautiful book-bindings are very fine and numerous.   The manuscript department has priceless treasures in many languages which illustrate the progress in writing from the second century before Christ to the fifteenth century.   There is a portion of the papyrus containing Aristotle's lost treatise "On the Constitution of Athens," and there is the Magna Carta of King John.   Among other rarities may be mentioned the earliest Greek text of the Scriptures on vellum, the Latin Bible of St Jerome, and the English Bible of Wyclif and his disciples.   The library, the manuscript department, the print room, and the newspaper room are the most frequented parts of the museum, and are visited for purposes of research by people from all parts of the world.

The ancient sculpture of the museum is superior to any other single collection in Europe, and affords a complete series from Egypt, Assyria, Greece and Rome. The Egyptian antiquities are the earliest examples of ancient sculpture, and in the large rooms there are sarcophagi, columns, tablets of the dead, and sepulchral urns. One of the most interesting relics is the Rosetta Stone, which contains a decree of the time of Ptolemy V, probably about 196 B.C. This celebrated stone furnished the first clue towards deciphering the ancient Egyptian hieroglyphics.

The Assyrian antiquities are of a most interesting character, and were brought from Nineveh by Mr Layard and others. The sculptured slabs represent the wars and conquests, the battles and sieges of the Assyrian monarchs; the colossal statues of human-headed lions and bulls give us some idea of the great conceptions of the artists.

The collection of Greek sculpture is the chief glory of the British Museum, for the Elgin Room contains the so-called "Elgin Marbles," unequalled works of Greek sculpture, executed without doubt by Pheidias and those of his school. They are from the Parthenon at Athens, a temple that was consecrated in 438 B.C., and still stands, a noble ruin, to this day.

The department of coins and medals is of great value, for it contains no less than 250,000 specimens, which are arranged vertically into historical compartments, and horizontally into geographical. We can thus gain a complete view of coins current throughout the civilised world during a particular period or century, and trace the development from the rudest to the finest art.

The gems and gold ornaments form a collection which is probably the richest of its kind in the world; and the Waddesdon Bequest Room has the collection of arms, jewels, plate, enamels, carvings, and other works of art, which were bequeathed to the museum by Baron Ferdinand Rothschild. It constitutes the finest bequest ever made to the museum, and is worth at least £300,000.

The collections of glass and pottery are very rich and valuable, and the prehistoric remains and Early British antiquities are most varied and extensive. The ethno-graphical gallery contains a collection of objects illustrating the habits, dress, warfare, handicrafts, and religions of the less civilised people of the world.

We will now give a brief survey of the Natural History Museum, which, as we have said, was built to contain the natural history collections of the British Museum. This enormous building, in the Romanesque style, was erected from the designs of Mr A. Waterhouse, and stands on the site of the International Exhibition of 1862. It is remarkable for the varied terra-cotta decora-tions on the external façades and internal wall surfaces. On the western side of the building the ornamentation is based on living organisms, while on the eastern side it is derived from extinct specimens.

As we enter, we find ourselves in the Great Hall, with its grand staircase facing us, surmounted by a statue of Darwin. In the Great Hall are exhibited the introductory series to the zoological and botanical collections, and illus-trations of general laws in natural history. These latter are of great interest, illustrating such principles as variation

under domestication, external variation according to sex
and season, albinism, melanism, and protective resemblance
and mimicry.

The gallery of British zoology contains a collection of
animals of all classes, which are, or have been in recent
times found, in the British Isles.   The collection of birds

Natural History Museum, South Kensington

is specially beautitul, not only in themselves, but for the
charmingly natural way in which they are set up.   Here
is the wonderful Gould collection of humming-birds, and
in a great gallery not open to the general public are
stored some hundreds ot thousands of specimens representing the birds of every portion of the globe.   The
first floor has stuffed mammals, and the series of gorillas,

chimpanzees, and orang-utans attract special attention. The mineral collection is very extensive, for it has specimens of every mineral species and variety.

The botany department has two divisions. In one, the collections consist of specimens suitable for exhibition, and are meant to illustrate the various groups of the vegetable kingdom and the chief facts on which the natural system of plant classification is based ; while the other, which comprises some splendid herbaria, is set apart for persons who are engaged in the scientific study of plants.

Occasionally special exhibits are on view in the museum, and during 1909, the centenary of the birth of Charles Darwin, there was a very notable collection of specimens, books, and prints illustrating the life-work of one who may be said to have wrought a greater change in scientific thought than any man since the day of Newton.

The buildings known officially as the Victoria and Albert Museum stand on some 12 acres of land purchased out of the surplus proceeds of the Exhibition of 1851. The idea of the original South Kensington Museum originated with the Prince Consort, and the first buildings were opened in 1857. These were of a temporary character but, as time went on, fine permanent buildings were erected. A great deal of the beautiful work in their decoration was carried out by an enthusiastic body of artists, and to them we owe the fine terra-cotta work on the north side of the quadrangle, the faïence in the refreshment rooms and staircase, and the pavements. Some of the tile-work was designed by Sir Edward

Poynter, and the firm of which William Morris was the head designed and carried out the charming decorations in the grill room. Subsequently the fine science school was built, and then the library and two great architectural courts were erected. The South Kensington Museum, as it was called down to 1909, played a most important part in the education of the British public, and largely to its influence we may attribute the more artistic taste now shown in the decoration and furnishing of our homes.

The accommodation for the exhibits was found to be far too limited, and in 1898 it was decided to prepare plans for a new building, the foundation stone of which was laid by Queen Victoria in 1899. From that year to 1909, the work went on, and then the Victoria and Albert Museum, as it is appropriately called, was opened by King Edward VII.

The new buildings are of a most attractive character, and are from the designs of Sir Aston Webb, R.A. The architect says that "a free Renaissance treatment has been adopted as the one that best admits of the introduction of the very large amount of window space required in such a building." The museum has a frontage to Cromwell Road, and also to Exhibition Road. The main entrance is in the former, and has a great portal finished by an opening lantern of the outline of an imperial crown, to mark its character as a great national building. The scheme of sculptured decoration on the front includes statues of 32 famous British artists and craftsmen; and there is a figure of Fame on the lantern, and two figures on the buttresses below, representing Sculpture and

Architecture. There are statues of Queen Victoria and the Prince Consort, and of King Edward VII and Queen Alexandra. The carved panels in the archivolt bear a quotation in letters of gold from Sir Joshua Reynolds's *Discourses*—" The excellence of every art must consist in the complete accomplishment of its purpose."

The internal arrangements of the new buildings are admirable. Although on simple lines, "an attempt has been made to prevent weariness to the visitors by avoiding galleries of undue length, by providing vistas and glimpses through the building in passing, and by varying the sizes, proportions, and design of the various courts and galleries." The new buildings double the accommodation for the exhibits, and when the contents are re-arranged, London will have one of the finest museums in the world.

The primary object of the museum is to provide models for, and to aid the improvement of, such crafts and manufactures as are associated with decorative design. The contents are grouped by industries, and the departments contain collections relating to woodwork, furniture, and leather; metal work; textiles; architecture and sculpture; engraving, illustration, and design; library and book production; paintings; ceramics, glass, and enamels. From this classification of the varied contents of the museum, it will be seen how comprehensive is its scope. We have space only to mention a few of the notable objects that are exhibited. The famous Raphael Cartoons bought by Cromwell, and hung at Hampton Court from the time of William III till they were transferred to this museum, are a great attraction. The relics of the old architecture of London are most interesting, especially the carved oak front of Sir Paul Pindar's house in Bishopsgate Street, and a room with fine oak panelling, an ornamented ceiling, and a stone fireplace, from Bromley Palace in the east of London. The collections have been much enriched by the beneficence of private donors, chief among whom were Mr Sheepshanks, who bequeathed his pictures, and Mr John Jones, whose gifts of French furniture, porcelain, bronzes, and other objects were valued at £250,000. In the southern galleries may be seen Stephenson's first locomotive the "Rocket," the first hydraulic press by Bramah, the engine of the "Comet" the first steamboat, and many models of ironclads, liners, and lighthouses.

The India Museum is really a branch of the old South Kensington Museum, and comprises objects formerly in the possession of the East India Company. There are examples of Hindu architecture, models of Indian divinities, and a rich collection of brocades and shawls from the gorgeous East. At present, the future of this museum is under consideration, and there is reason to believe that it will be merged in the general museum.

The Imperial Institute was erected as a memorial of Queen Victoria's Jubilee, and was opened in 1893. The building is one of the finest in London, and the chief feature, the central tower, rises to a height of 300 feet above the great portal. The prevailing style of the building is a free rendering of the Renaissance, which has been so much in vogue during the last 20 years. The Institute has been reorganised, as it did not quite fulfil the intentions

of the founders, and part of the building now belongs to
the University of London, the remainder being under the
control of the Board of Trade. The object of the Institute

The Imperial Institute

is to organise and illustrate the industrial and commercial
resources of the Empire, and the collections contain products of our different Colonies and Dependencies.

25. Public Buildings—(*d*) Art Galleries. National Gallery, National Portrait Gallery, National Gallery of British Art, The Wallace Collection.

The art collections of London are unrivalled for their variety and beauty, and the numerous national galleries are now well arranged and much appreciated not only by connoisseurs but also by the general public. The collections of pictures and sculpture in the palaces and in some of the mansions of the wealthy are magnificent, but as they are not open to the public we can only record the fact of their existence. The most popular of the national collections are at the National Gallery, the National Portrait Gallery, the Tate Gallery, and Hertford House. We have already noticed the contents of the Victoria and Albert Museum, so that we need not further deal with the treasures of that fine institution.

From an architectural point of view the National Gallery is the least satisfactory of all our national buildings. It occupies an excellent position on the north side of Trafalgar Square, which has been called the finest site in Europe, but its elevation is not imposing, and the dome and the cupolas are mean and seem to dwarf the building. The architect was not allowed a free hand in his designs, so that some allowance must be made for him. The building is in the Classical style, raised high upon a terrace; it was opened in 1838. Mr Ruskin,

who derides the dome—"such as it is," are his words—
is bound to confess that the National Gallery is "without
question the most important collection of paintings in
Europe."

The arrangement of the pictures is according to
schools, with a close adherence to chronological order.
With very few exceptions, all the schools are represented,

The National Gallery and St Martin's Church

and every year sees an accession to the treasures of this
gallery, either by purchase or by the generosity of art
lovers. The Umbrian School is one of the most interesting departments, and here is the "Ansidei Madonna"
by Raphael. This celebrated picture, which is by many
considered one of the finest pictures in the world, was
bought from the Duke of Marlborough for £70,000. In
the Venetian room the notable pictures are very numerous.

The Dutch pictures, with their realistic and domestic atmosphere, are very different from those of the Italian masters. The Dutch interiors, the landscapes, and still life are excellent. A most famous picture is Van Dyck's "Charles the First." The pictures of the Spanish school, too, are magnificent.

We have not space to glance at many of the other schools so well represented in the National Gallery, but we must emphasise the value of the splendid collections representing British art. Here we have the great works of Sir Joshua Reynolds illustrated by his portraits; of Hogarth "that great moralist and satirist in paint"; of Gainsborough and Constable distinguished in portraiture and landscape; and of such notable artists as Wilkie, Opie, Raeburn, Crome, Morland, and Landseer.

The National Portrait Gallery adjoins the National Gallery, but is in a separate building which was opened in 1896. It contains a most interesting collection of the worthies and notabilities of the English race, and its portraits have been called a "graphic dictionary of national biography." This National Roll of Honour contains upwards of 1000 portraits and covers the whole period of our history from the reign of Richard II. As far as possible they are arranged in chronological order, but, with the exception of the reigning sovereign, living personages are not represented.

The National Gallery of British Art, or the Tate Gallery as it is familiarly called, is a branch of the National Gallery. It is at Millbank and was founded by Sir Henry Tate in 1897. The building stands on

the site of Millbank Prison, and is eminently satisfactory as an art gallery. Its chief exterior feature is
the Corinthian portico supporting a figure of Britannia.
The galleries are built around a central hall which rises
into a dome. The rooms are well lighted, and the
pictures admirably hung and not overcrowded. The
Tate Gallery has the collection of 65 pictures given

The Tate Gallery

by Sir Henry Tate, and the pictures and sculptures
bought under the Chantrey Bequest. The Turner
collection of pictures bequeathed to the nation by that
great artist, and representative of his earlier and of his
later manner, has been transferred from the National
Gallery to the Tate Gallery. There are also examples
of pictures of the British school, mainly of the mid-
Victorian period, and more especially of Millais, Holman

Hunt, Madox Brown, Burne-Jones, and of the rest of the small band of artists who adopted the name of "Pre-Raphaelite." The finest collection is, however, in the rooms which contain the chief works of George Frederick Watts, one of the greatest painters of the Victorian Age. The pictures were presented to the nation by the artist, and convey some of the lessons he was anxious to teach to the men and women of our time.

We now pass to the Wallace Collection in Hertford House, Manchester Square, which once belonged to the Marquis of Hertford, but became the property of the British nation in 1897. It was probably the finest private collection in the world, and its value has been estimated at £4,000,000. The special charm of the Wallace Collection is that the pictures and other works of art remain to a large extent as the skill and taste of their former owners had placed them. Hence they seem rather as the ornaments of a fine mansion than as so many items in a museum or picture gallery. The collection consists of arms and armour; furniture and other objects of art; and pictures. The collection of armour is unique; the French furniture is unrivalled; and the Sevres porcelain is comparable only to that at Buckingham Palace and Windsor Castle. In closing this brief reference to the Wallace Collection, we may add that the pictures representing the French School of the eighteenth century are equal to those in the Louvre.

26. Public Buildings—(e) The Hospitals. St Thomas's, St George's, Charing Cross, Royal Military Hospital, Foundling Hospital.

London is proud of its hospitals, and their administration is improving year by year. In no other city is so much done to alleviate pain and suffering. The subscriptions of all classes for the maintenance of the hospitals amount to a very large sum each year. There are three great funds which raise money for this purpose; first there is King Edward's Hospital Fund, which was inaugurated in the year of Queen Victoria's Diamond Jubilee; and then there are the Hospital Sunday Fund, and the Hospital Saturday Fund. Many of the large hospitals are without medical schools, but some of the oldest and best have them as part of the foundation. The total income of all the London hospitals in 1910 amounted to nearly £800,000, and in the same time about 4,000,000 attendances were made by out-patients, while nearly 93,000 in-patients were accommodated. Besides the large general hospitals, there are special hospitals for the treatment of children, as well as for consumption, skin disease, fever, and cancer.

In the volume on East London special reference was made to St Bartholomew's, Guy's, and the London Hospitals, therefore in this volume we shall notice only St Thomas's, St George's, Charing Cross, the Foundling, and Chelsea Hospitals.

St Thomas's Hospital stands on a fine site on the right bank of the Thames facing the Houses of Parliament. It is one of the most ancient, and also one of the richest and largest of the London Hospitals. Founded about 1213, it was at first an almshouse endowed by the Priory of Bermondsey. As time passed, its usefulness increased, and after many changes it was removed in 1871 from its

St Thomas's Hospital

original site in Southwark to its present riverside frontage. The buildings are of red brick relieved with Portland stone, and in front of them runs the Albert Embankment. The yearly income is over £63,000, and is mainly derived from invested property. There are 561 beds, and the staff consists of 126 nurses. In 1910 there were 7316 in-patients, while 85,581 out-patients were given advice and treatment. St Thomas's has a famous medical school, and it holds a high place in the history of nursing, for the Nightingale Fund Training School for Nurses is associated with it.

St George's Hospital opposite Hyde Park Corner was founded in 1733, but it has only occupied its present position since 1829. It has a medical school, and the nursing staff numbers 150. There are 334 beds, and in 1910 they were occupied by 4867 patients. During the same period 48,583 out-patients visited the hospital. The most distinguished name associated with this hospital is that of John Hunter, the great anatomist, who died suddenly in the old building in 1793.

Charing Cross Hospital founded in 1820 has an annual income of about £23,000, a very small portion of which is derived from investments. The present building was designed by Decimus Burton in 1831, and much enlarged in 1904. It has a medical school, a staff of 64 nurses, and 150 beds. During 1910 there were 2112 in-patients, and 21,863 out-patients.

We now pass to the consideration of the Royal Hospital at Chelsea, and of the Foundling Hospital in St Pancras. The Royal Military Hospital stands at a little distance from the river towards the east of Chelsea, and is devoted to the accommodation of old disabled soldiers. The first stone of the building, which was designed by Sir Christopher Wren, was laid by Charles II in 1682, but it was not completed till 1690. The structure is of red brick faced with stone, and is a good specimen of Wren's work. It is well-proportioned, and is built round three courts. The chief façade faces the river, and in the centre stands a bronze statue of Charles II.

In the Great Hall there are portraits of great commanders, and the remnants of flags that have been carried on many a battlefield. There are about 540 pensioners

Chelsea Hospital: the Dining Hall

in this Hospital, and every week they assemble in the Hall to receive their pay. This weekly muster is a picturesque sight, for the men in their uniform, which consists of red coats in summer, and blue coats in winter, with quaint peaked caps, or cocked hats when in full dress, represent the élite of our veteran soldiers. The men have little to do, and all reasonable liberty is allowed

them. They are expected to attend service on Sunday
in the chapel, and such a scene has been portrayed by
Sir Hubert Herkomer in his celebrated picture "The
Last Muster." The remains of the Duke of Wellington
lay in state in the Great Hall in November, 1852. The
Hall was draped in black and lighted with wax tapers,
and crowds of people were admitted to this solemn
spectacle. Around the Hall stood picked soldiers of the
Grenadier Guards with their arms reversed; while Yeo-
men of the Guard were on duty around the bier.

There is an enclosure of about 13 acres to the north
of the Hospital, and this is planted with limes and horse
chestnuts, while towards the south are extensive gardens.
Altogether the Hospital and grounds occupy 50 acres.

The Foundling Hospital has a unique place among
the many charitable institutions of London. It has a
curious history, for it was founded in 1739 by Captain
Thomas Coram, for exposed and deserted children. The
founder was a benevolent seaman whose heart was
touched by the deserted infants so often found in the
streets of London. When first instituted, the hospital
was open to all children, whose mothers had simply to
leave them in a basket placed for the purpose at the
gates. After a while other regulations were made, and
now 500 children are admitted. From its early days
this charity has attracted the sympathy of many great
men. Handel gave the organ to the chapel, and there
he often presided at performances of his *Messiah* for the
benefit of the institution. Hogarth painted the portrait
of the founder which hangs in the board-room, and

The Foundling Hospital : the Chapel

(From an old print)

Dickens describes the place in *Little Dorrit*. The building,
of brick with stone dressings, has a centre and wings,
with spacious gardens behind and a playground in front,
where the boys in red waistcoats and the girls in their
white aprons may be seen at their games.   On Sundays
the services in the chapel are frequented by many visitors,
both for the trained singing of the children and their
picturesque appearance.   The boys and girls sit on each
side of the famous Handel organ; the boys in red sashes
and the girls in white mob caps, tuckers, and aprons.
The altar-piece in the chapel is a fine picture, "Christ
Blessing Little Children," by West, and the Hospital
has also among its treasures Hogarth's "Finding of
Moses," and Raphael's cartoon of the "Massacre of the
Innocents."

27. Education—Primary, Secondary, and Technical. Foundation and Collegiate Schools. The University of London.

Before the year 1870, the elementary education of
the children of London was not compulsory, and was
managed by the Church of England and other religious
denominations.   Mr W. E. Forster introduced a bill for
the compulsory attendance of all children at school, and
when this Education Act was passed in 1870, a body
known as the London School Board was formed.   It
consisted of 55 members, and for a period of 33 years

was the directing authority for much of the elementary education in London. During its *régime* the Church, Catholic, Wesleyan, and other denominational schools were controlled by their own managers, and had nothing to do with the London School Board.

The work of the London School Board came to an end in 1903, when Mr Balfour passed a new Education Act by which the London County Council became the Education Authority for the County of London. The London County Council actually superseded the London School Board on May 2, 1904, and an Education Committee was then formed to deal with all classes of education. At the present time the Education Committee consists of 40 members of the London County Council together with 12 co-opted members, who are specially chosen for their interest in the work of education.

The great merit of the last Education Act is due to the co-ordination of all branches of education in the hands of one body. Thus the London County Council have charge of all the elementary schools, both those belonging to the late London School Board, and the denominational schools. The latter schools, however, are still allowed to give their own religious instruction, and their managers have some control over the teachers in these schools.

Now in considering the extensive duties of the London Education Committee, we will begin with elementary education. There are about 920 schools for this purpose, and they have accommodation for over 750,000 scholars, whose ages vary from three to 15. The children have a sound elementary education, which is well graded for their capacities. There are also many special schools for instruction in such subjects as cookery, laundry-work, housewifery, and manual work, and also for the separate treatment of children who are deaf, blind, and mentally and physically defective.

The Council have also a great many departments known as Central Schools, where the curriculum is of a more advanced character, and the course of instruction is arranged for four years, after an entrance examination. For the whole of the elementary schools there are about 20,000 teachers employed, and altogether the London County Council expend upwards of three million pounds yearly on its educational work.

The Higher Education work of the London County Council began by taking over the duties of the Technical Board, and since 1904 it has been concerned with technical, secondary, and university education. The Council has adopted the policy of maintaining and developing the work of existing institutions in London before erecting new institutions under its own management, and to give facilities for scientific and technological instruction in every district of London. The most important of all the institutions which provide technical education are the 12 Polytechnics, where instruction is given in the ordinary branches of science and art, as well as in the engineering, building, and chemical trades. There are other institutes which are specially devoted to the teaching of one particular craft, and these are styled Monotechnic Institutions. Thus there is one school devoted to training craftsmen in photo-process work and lithography, a second to carriage-building, and a third to leather-tanning and leather-dyeing.

The London County Council has numerous Schools of Art under its control, and there is also the Central School of Arts and Crafts, which provides for the artisans of London instruction in decorative design. The trades for which provision has been specially made at this school are those directly or indirectly associated with the building trades, such as decorators, stone-carvers, metal-workers, cabinet-makers, and designers of wall-papers.

Considerable grants are made by the London County Council to the University of London, and to University College, King's College, Bedford College, and the London School of Economics, and in return these bodies give a certain number of free places to the nominees of the Council. The London County Council spend large sums of money in the award of scholarships which carry pupils from the elementary to secondary schools. It has now 16 secondary schools of its own, and it further makes annual grants to other schools that receive its scholarship holders.

Besides the schools under the control of the London County Council there are also some great Public Schools which must be specially mentioned. First there is St Paul's School, the most ancient, for it was founded in 1509 by Colet, Dean of St Paul's. Colet was one of the leaders of the Renaissance in England and St Paul's was the first English school in which Greek was publicly taught. For many years it was under the shadow of the

great Cathedral, but in 1884 it was removed to its present
fine buildings in the Hammersmith Road. Westminster
School is at the back of Westminster Abbey; it was
founded by Henry VIII out of the spoils of the monasteries, and richly endowed by Elizabeth.

Two other famous public schools formerly stood in
the very heart of the City. Christ's Hospital—or as it was
commonly called the "Blue Coat School"—was one of the

St Paul's School

most cherished institutions of London. It was founded
in the reign of Edward VI, and its scholars still wear
the picturesque dress of that period. Owing to want of
room, the school was removed to a spacious site near
Horsham in Sussex, and the old buildings have given
place to the extension of the General Post Office. The
Charterhouse School was removed from London in 1872,
and was established in a fine building at Godalming

in Surrey. The Charterhouse was founded in 1611 by
Thomas Sutton, and though the boys have gone, the
brethren of the foundation, some eighty in number, live
on in the same place, in collegiate style. Merchant
Taylors' School, founded by the Merchant Taylors'
Company, dates back to 1561. In addition to these great

University College

and famous schools of the past, there are others of a
later date, which give a similar education. Among these
may be mentioned Dulwich College, the City of London
School, University College School, and King's College
School.

We now come to the last section of this chapter,
which has to deal with the University of London.

Founded in 1836, the University was for many years an examining body, and had nothing to do with the work of teaching. Its examinations proved whether students had been well taught in certain subjects, and whether they merited its certificates or degrees. In 1898, however, the University became a teaching body as well as an examining board. Its headquarters were formerly at Burlington House, Piccadilly, but after its re-organisation, the central block and one of the main wings of the Imperial Institute at South Kensington were assigned for this purpose. The various colleges, and medical schools, such as King's College, University College, Bedford College, and others are now "Schools of the University." The number of students attending the 42 Schools of the University are about 3500, who pursue a course of study approved by the University.

28. Roll of Honour.

For some years before 1901 the Society of Arts placed tablets on houses of historical interest in London, and since that date the London County Council have undertaken this work. There are now more than one hundred houses that are so indicated, but the work is by no means completed, for there are many celebrated men who have lived and died in London whose houses have not yet been marked by a memorial tablet. In addition to this commemoration of noted Londoners there are also statues to some of them in the streets and squares, as well as monuments to them in the cathedrals and churches of the metropolis. In this rapid survey of men who have conferred distinction on London it will be possible to do little more than just mention the locality with which they were connected, or the special work which links them to the great City. The Roll of Honour of London necessarily includes men whose names are of world-wide significance, and whose memory is honoured by other towns and counties in our country; but there are many of them who are indissolubly linked with the associations of London. To use the words of Lord Rosebery it may be said that in taking a walk in London "it is an immense relief to the eye and to the thoughts to come on some tablet which suggests a new train of thought, which may call to your mind the career of some distinguished person, and which takes off the intolerable pressure of the monotony of endless streets."

We need not spend much time on the royal personages who are associated more particularly with London. The fact that it has been the capital for nearly a thousand years, and that the Court has resided there for the greater part of that period, tells us at once that all our monarchs have some claim, either by birth or residence, to be considered Londoners. There are a few, however, that we recall at once for some special reasons. Alfred has been called the founder of London ; William I built the Tower for its protection; and Charles I was beheaded at White-hall. Westminster Abbey with all its historic associations has been the crowning place of our sovereigns, and there, too, many of them are buried.

Neither will it be necessary to recount the long list of divines who have spent much of their time in London. When we remember that St Paul's Cathedral and Westminster Abbey have given us a succession of noted bishops and deans, and that Lambeth Palace, the official residence of the Archbishop of Canterbury, has been the home of nearly one hundred successors of St Augustine, we at once realise what a part London has played in the religious life of our nation. Here also, however, we may select a few outstanding names. Becket, son of a wealthy merchant, in many ways the most famous of our archbishops, was born in London, behind the Mercers' Chapel in the Poultry. John Colet, Dean of St Paul's, friend of Erasmus, was the founder of St Paul's School. John Wesley was educated at the Charterhouse, and held his first Methodist Conference in London in 1744. Sydney Smith, the witty Canon of St Paul's, had previously been preacher at the Foundling Hospital and at Berkeley Chapel. Frederick Denison Maurice, the leader of the Broad Church movement in the reign of Victoria, was Chaplain at Guy's Hospital and Lincoln's Inn, and one of the founders of the Working Men's College, which did so much for the intellectual improvement of the artisan class. John Henry Newman was born in London in 1801, and lived near Bloomsbury Square, in the garden of which he and young Disraeli used to play. Outside Brompton Oratory and facing Brompton Road there is a statue of Newman as Cardinal.

The statesmen who have identified their fortunes with London are very numerous. Thomas Cromwell was born at Putney, and, after helping Henry VIII in the dissolution of the monasteries, was beheaded on Tower Hill. Walpole, the great prime minister of the early Georgian period, lived in Arlington Street. Edmund

John Henry, Cardinal Newman

Burke received his legal training at the Middle Temple, and his speech in Westminster Hall on the impeachment of Warren Hastings was one of his greatest efforts. He was fond of London, and among his literary friends were Dr Johnson and Goldsmith. The great Earl of Chatham was born in the parish of St James, Westminster, and his son, William Pitt, one of our foremost prime ministers, has a statue to his memory in the Guildhall. George Canning, who followed in Pitt's footsteps, was brought up in London, and trained at Lincoln's Inn. His statue faces the Houses of Parliament. Sir Robert Peel has special claims on our notice, for not only did he repeal the Corn Laws, but he formed the Metropolitan Police Force—or " Peelers " as they were at first termed—who took the place of the old watchmen. He was thrown from his horse near Hyde Park Corner, and died from the effects of the fall at his house in Whitehall. Peel's services to London are brought to our mind by his statue at the west end of Cheapside. Benjamin Disraeli, afterwards Lord Beaconsfield, became one of the most interesting prime ministers of the Victorian era. He was born in Theobald's Road, and died at 19, Curzon Street, Mayfair. His statue by Boehm is in front of the Houses of Parliament where he won his triumphs. Gladstone, too, spent much of his time in London. He lived in Harley Street and in Carlton Terrace, and the statue to his memory in the Strand is one of the most striking in London. Gladstone was buried in Westminster Abbey, and there his statue is near that of his great rival, Lord Beaconsfield.

Among the men of action whose fortunes were connected with London, we will name the two foremost in our history. Nelson, the greatest admiral since the world began, and Wellington, the hero of a hundred fights, are assuredly the pride and possession of the Empire ; and

London honoured them both with public funerals which were unique in their pomp. Both of them were laid to rest in St Paul's, where there are splendid monuments to their memory. Nelson's monument in Trafalgar

Lord Beaconsfield

Square is one of the sights of London, and everyone knows the equestrian statue of the Duke of Wellington in front of the Royal Exchange. The Duke lived at Apsley House, Piccadilly, and when he was out of favour with the London mob, the windows of that residence were broken. General Gordon was born at Woolwich, and has a statue in Trafalgar Square. Here we may mention that London has been very generous in raising statues to its military heroes, and Havelock and Napier, the Crimean soldiers, and those who fought in South Africa and elsewhere, are all commemorated either in the public squares or in the cathedrals and churches of the metropolis.

The historians and antiquaries who flourished in London form a goodly company. Leland, who was born in London about 1506, was educated at St Paul's School. Among his works *The Itinerary* is the best known, and shows that even in early Tudor days men were beginning to take an interest in the past history of their country. John Stow, who formed a worthy successor to Leland, was the son of a tailor of Cornhill, where he was born in 1525. Stow was the first historian of London, and his *Survey of London and Westminster* is the foundation of all later work on that subject. A very striking monument to his memory is in the church of St Andrew Undershaft. William Camden, scholar, antiquary, and historian, was educated at Christ's Hospital, and St Paul's School. He stands out as the great historian of his country in the reign of Elizabeth, and his *Britannia* does for the whole country what Stow did for London. John Strype, who lived in the later Stuart period, received his education at St Paul's School. He is connected with Hackney, and continued Stow's *Survey of London and Westminster* to the beginning of the eighteenth century. Edward Gibbon, the author of the *Decline and Fall of the Roman Empire*, was born at Putney. In many ways that history is one of the greatest works in our literature, and we like to think of the author as the friend of Johnson, Burke, and Goldsmith. Henry Hallam, the author of the

Edward Gibbon

Constitutional History of England, and other historical works, lived for 20 years at 67, Wimpole Street. That house has a singularly pathetic interest, for it was also the home of his son, Arthur Henry Hallam, Tennyson's dear friend, to whose memory *In Memoriam* was the poet's tribute. The following lines from that poem refer to the house :—

> " Dark house, by which once more I stand,
>   Here in the long unlovely street,
>   Doors, where my heart was used to beat
> So quickly, waiting for a hand,
>
> .   .   .   .
>
> A hand that can be clasped no more."

George Grote, whose *History of Greece* was his chief work, died at 12, Savile Row. Lord Macaulay, one of our most picturesque and graphic historians, spent many happy years at the Albany in Piccadilly, removing afterwards to Holly Lodge, Campden Hill, where he died in 1859. His worth as the writer of the *History of England* and as an essayist are recognised by his statue in Westminster Abbey, where he was buried.

London has always been famous for its poets, and during the Elizabethan period it was "a nest of singing-birds." One of our earliest poets, John Gower—" Moral " Gower as he was called—lies buried in Southwark Cathedral, where there is an effigy to his memory. Chaucer, too, has made Southwark famous for all time. He was born in London, resided in Aldgate, and became Comptroller of the Petty Customs. It was at the Tabard Inn, Southwark, that his company of pilgrims assembled for their journey to Canterbury. Chaucer subsequently lived in Westminster, and was buried in the Abbey. Edmund Spenser, author of the *Faerie Queene*, was born in London, probably near the Tower. He ended his life in distressed circumstances at an inn in King Street,

Westminster. Shakespeare, the greatest of all our dramatists, came to London when he reached manhood, and his first work there was probably some lowly office in connection with the Curtain Theatre at Shoreditch. All his work in London was either as actor or playwright,

Edmund Spenser

and the Globe and the Blackfriars Theatres on the Surrey side were the scenes of his triumphs. His contemporaries in London were Beaumont and Fletcher, Massinger and Ben Jonson, and the latter he often met at the Mermaid Tavern in Bread Street. His friend,

Ben Jonson, realised his greatness, for he says that Shakespeare wrote "not for an age but for all time."   There
is a fine monument to Shakespeare in the Abbey, and
a later one in Leicester Square.   John Milton, the greatest

Ben Jonson

of the Stuart poets, the son of a London scrivener, was
born in Bread Street, Cheapside.   He was educated at
St Paul's School, and lies buried in St Giles's, Cripplegate,
where there is a monument to him in the churchyard.
Among the London poets of the early Georgian period

Dryden, Pope, and Gray take a high place, but Goldsmith has stronger claims on our attention. He reached London in destitution, became a physician in Southwark, and then usher in a school at Peckham. Eventually he found the friendship of Dr Johnson, and became a member of the famous Literary Club. He died at 2, Brick Court,

John Milton

Temple, and was buried in the Temple Church. William Blake, poet and artist, was born in 1757 in Broad Street, Golden Square. He had many residences in London, and died in Fountain Court, Strand, on 12 August, 1827. Samuel Rogers, Lord Byron, and Keats were three London poets who helped to make the early nineteenth century

famous. Lord Byron was born in London, at 16 Holles
Street, Cavendish Square, and there is a statue to him in
Hyde Park ; Rogers died at 22, St James's Place ; and
Keats, in many ways the most brilliant of the trio, was
born at Moorfields, and passed much of his short life at

Alexander Pope

Well Walk, Hampstead. Of the Victorian poets, Robert
Browning was born at Camberwell, educated at Peckham
and at University College, and buried in Westminster
Abbey ; Thomas Hood, who wrote *The Bridge of Sighs*

and the *Song of the Shirt*, was born in the Poultry quite close to Bow Church; and Dante Gabriel Rossetti, one of a gifted family, was born at 28, Charlotte Street, now 110, Hallam Street.

William Blake

The men of letters who have made London their home are even more numerous than the poets. Caxton, the first English printer, set up his printing press at Westminster, where he published eighty separate books, the first being *The Dictes and Sayings of the Philosophers.*

Pepys, writer of the famous *Diary*, was educated at St Paul's School and lived in Buckingham Street, Strand. He was buried in St Olave's, Hart Street, at nine o'clock at night ; and there are monuments in that church to him

Samuel Pepys

and to his wife. For a correct and realistic knowledge of London of the time of Charles II his book is without a rival. Evelyn, too, who wrote a *Diary* of the same period, lived at Deptford, and his work though less lively, is of high value. Nor must we forget Daniel

Defoe, who wrote the *Journal of the Plague Year*. He was born in St Giles's, Cripplegate, 1661, and is buried in Bunhill Fields Cemetery. Of all the men of letters of whom London is justly proud, none is greater than Dr Johnson. He is styled the "leader of literature in the eighteenth century," and the best part of his life's work was accomplished in London. The celebrated Club founded by Sir Joshua Reynolds at the Turk's Head, Gerrard Street, Soho, included among its members Johnson, Burke, Goldsmith, Gibbon, and Boswell, but Johnson was the acknowledged leader. London was the greatest place in the world to Johnson. "Fleet Street," he once said, "has a very animated appearance; but I think the full tide of human existence is at Charing Cross." The house in Gough Square, where he wrote the *Dictionary*, still stands; his seat in St Clement Danes Church in the Strand has a brass plate on it; and the *Cheshire Cheese*, one of his favourite haunts in Fleet Street, is yearly visited by thousands of his admirers. Johnson died at 8, Bolt Court, Fleet Street, and was buried in Westminster Abbey. Happily, we can follow Johnson in his London life, for Boswell's biography of his hero gives us the minutest details of his friends and their homes. Charles Lamb, the gentle essayist, was a true Londoner. Born in the Middle Temple he was educated at Christ's Hospital, and was a clerk in the East India House for 36 years. Most of his essays were written in London before he retired to Enfield. When we come to the Victorian era, we have a goodly company of men of letters who delighted in London. Thackeray, the great novelist, was educated at the Charterhouse, and *Vanity Fair*, *Esmond*, and *Pendennis* were written at Young Street, Kensington, or Kensington Palace Road, where he died. Dickens, too, was essentially a Londoner. He knew the metropolis of his time as few others knew it, and in such novels as *David Copperfield*, *Sketches by Boz*, and the *Pickwick Papers* we find evidences of this intimate knowledge. He lived at 48, Doughty Street, and in the Marylebone road. Thomas Carlyle, essayist, historian, and man of letters, lived at 5, Cheyne Row, now 24, Cheyne Row, for 50 years. There he wrote most of his books, and his house is now a museum, visited by thousands every year. Leigh Hunt, the essayist and author of *The Town*, lived at Upper Cheyne Row, Chelsea, and afterwards at Hampstead. James Mill and his son John Stuart Mill lived in London, the latter being born in Rodney Street, Pentonville. With "George Eliot," who wrote *The Mill on the Floss* at Holly Lodge, Wimbledon Park Road, and John Ruskin, the fine-art critic and brilliant prose-writer, who lived for a while at Croydon, we must close our review of the Victorian men of letters.

When we turn to the men of science who lived in London, we find such names as Newton, Herschel, Hunter, Jenner, Darwin, Lyell, Faraday, Huxley, and Lord Kelvin. Sir Isaac Newton lived at 87, Jermyn Street, and was Master of the Mint. Sir John Herschel, the great astronomer, lived at 56, Devonshire Street. He, too, was Master of the Mint, and was buried in

Westminster Abbey.   John Hunter, the eminent surgeon
and anatomist, began to practise at Golden Square, and
afterwards lived in Jermyn Street.   Among his pupils
were such men as Dr Abernethy, and Jenner, who

John Ruskin

introduced vaccination.   Darwin lived for some years in
Gower Street, and it was there that he wrote one of his
earliest books, the *Structure and Distribution of Coral
Reefs*.   Sir C. Lyell, the eminent geologist, who lived at
73, Harley Street, Faraday the chemist, and Huxley,

one of the greatest leaders of modern scientific thought, used to delight distinguished London audiences by their brilliant expositions of science. Lord Kelvin spent the closing years of a brilliant career at 15, Eaton Place. He

Michael Faraday

was buried on December 23, 1907, in the nave of Westminster Abbey, next to the grave of Sir Isaac Newton.

London has been the home of some of our greatest painters. Hogarth, born in Bartholomew Close, Smithfield, and apprenticed to an engraver in Cranbourne

Street, knew all the phases of London life and reflected them in his pictures. Sir Joshua Reynolds, the greatest English portrait painter and the founder of the Royal Academy, was the friend of Johnson and the great literary

Joseph Mallord William Turner

men of his time. Reynolds died at his house in Leicester Fields. Turner, our greatest landscape painter, was the son of a London barber, and Sir David Wilkie, whose paintings of humble life are so familiar, lived in about a dozen houses in various parts of London. John

Leech, a humorous artist, was born at 28, Bennett Street, Stamford Street, S.E., and was educated at the Charterhouse, where he was a fellow pupil of Thackeray. He died at Kensington, 1864. Lord Leighton, who may

John Leech

be called the last of the great Victorian artists, lived at Holland Park. There he built himself a house worthy of an artist, and there he died in 1896. Leighton House was presented to the nation after the death of the artist, to whom it is intended as a memorial. It is very beautiful and contains a large collection of his pictures and studies, and exhibitions are occasionally held in its rooms.

Our survey of London's famous men must close with a reference to three philanthropists. William Wilberforce, the man who devoted his life to freeing the slaves, lived at Broomwood House, Clapham; John Howard, the reformer of prison life, passed some of his days at 23, Great Ormond Street; and Sir Rowland Hill, who introduced penny postage, lived at 1, Orme Square, Bayswater. His statue stands at the east end of the Royal Exchange.

From this brief survey of London's Roll of Honour, it will be seen that the City has always had a charm, almost a fascination, over the lives of many of our great men. An attachment for London is the experience of most people who come to it early enough, and Dr Johnson expressed this feeling when he said :—"Why, sir, you find no man at all intellectual who is willing to leave London. No, sir, when a man is tired of London he is tired of life, for there is in London all that life can afford."

29. THE CITY OF WESTMINSTER AND THE BOROUGHS IN THE NORTH-WEST AND SOUTH-WEST OF THE COUNTY OF LONDON.

The City of Westminster comprises the following civil parishes:—Close of the Collegiate Church of St Peter, Liberty of the Rolls, Precinct of the Savoy, St Anne within the Liberty of Westminster, St Clement Danes, St George Hanover Square, St James Westminster, St Margaret and St John, St Martin-in-the-Fields, St Mary-le-Strand, and St Paul, Covent Garden.

Among the London boroughs it ranks thirteenth in point of population, and tenth with regard to area. The population is decreasing, and the density is 64 persons to the acre. There are 723 acres of open spaces, including the finest parks in London, namely, Hyde Park, part of Kensington Gardens, Green Park, and St James's Park. It is interesting to note that Westminster was a city for a brief period in the reign of Henry VIII. Edward VI, however, dissolved the bishopric, and to the end of the nineteenth century Westminster had no municipal authority. In 1899 it became one of the London boroughs, and in the following year it was created a city by royal charter. Although there is some dispute as to the origin of the name, most authorities are now agreed that when the monastery of St Peter was founded here, it was called the West Minster to indicate that it lay to the west of the East Minster in the City of London. It will also be remembered that this name gradually supplanted that of Thorney. Westminster has the royal palaces, and is the seat of government, containing as it does the Houses of Parliament, the Government Offices, and the Royal Courts of Justice. There are 14 wards in Westminster, and the borough council consists of 10 aldermen and 60 councillors.

Battersea comprises the area of the parish of Battersea, and its large population shows a density of 78 persons to the acre. In open spaces, Battersea is fortunate in having Battersea Park and large parts of Clapham Common and Wandsworth Common, comprising in all about 400 acres, or nearly one-fifth of the whole borough. The borough is divided into nine wards, and the council consists of nine alderman and 54 councillors. The municipal buildings on Lavender Hill are among the finest in London. The name, Battersea, has undergone several changes. In the Domesday Book it is called *Patricesy*, and has since been written Battrichsey, Battersey, and Battersea. The manor of Battersea belonged from a very early period to the Abbey of St Peter at Westminster, but passed to the Crown at the dissolution of the religious houses.

Chelsea is one of the smallest boroughs in London, its area being little more than one square mile. It is thickly populated, and the density of persons to the acre is 100. The borough is divided into five wards, and the council consists of six aldermen and 36 councillors. In a Saxon charter of Edward the Confessor, the name is written *Cealchylle*, and in Domesday Book it appears as *Cerechede*, and *Chelced*. At a later date we find Sir Thomas More writing it as *Cheleith*, and the present name, *Chelsea*, is probably derived from the nature of the place, for, says an old writer, "its strand is like the chesel which the sea casteth up of sand and pebble stones, thereof called Cheselsey, briefly Chelsey, as is Chelsey in Sussex." Chelsea was at one time a very aristocratic district, and was the residence of Sir Thomas More, Sir Hans Sloane, and other men of note. It was also famous for the Ranelagh and Cremorne Gardens, the Bun House, and for its china. The Physic Garden, formerly the property of the Apothecaries' Society, is in this borough, and Chelsea Hospital for Soldiers, built by Wren, is one of the chief buildings. Among the modern improvements are the Albert Suspension Bridge and the Embankment. Cheyne Row is associated with the name of Thomas Carlyle, who resided there for many years, as also did Turner, the famous landscape painter. Kalm, who visited England in 1748, relates that "Chelsea is almost entirely devoted to nursery and vegetable gardens," and was visited by "those who lived in London, now and then, especially on Saturday afternoons, for the fresh air, and the advantage of tasting the pleasures of a country life."

Fulham is below the average of the London boroughs in area, and its population shows a density of 90 to the acre. There are only five small open spaces in Fulham covering an area of about 68 acres. The borough is divided into eight wards, and the council consists of six aldermen and 36 councillors. From long before the Conquest, the manor of Fulham has belonged to the see of London, and Fulham Palace has been for three centuries the summer residence of the Bishops of London. The building is of no great antiquity, but the house and grounds are surrounded by a moat. The parish church of All Saints stands near the river, and has a fine tower 95 feet high, with a peal of 10 bells. The church is Perpendicular in style and has some good monuments. Fulham has always been famous for its nurseries and market-gardens; and about 1753 a manufactory of Gobelin tapestry was established here by Peter Parisot. It produced beautiful fabrics but the manufacture soon declined. Kalm, to whom earlier reference has been made, thus writes of Fulham in 1748:—"In appearance it is a pretty town with several smooth streets. All the houses are of brick, very beautifully built, some of which belong to gentlemen. Round about this place the country is full of gardens, orchards, and market-gardens, both for pleasure and use, and it can indeed be said that the country here is everywhere nothing but a garden and pleasance."

Hammersmith is the most westerly of the London boroughs. Its area is about three and a half square miles, and its population shows a density of 53 persons to the acre. The borough is divided into seven wards, and the council consists of six aldermen and 36 councillors. The handsome Town Hall in the Broadway was opened in 1897. Hammersmith Suspension Bridge across the Thames, erected in 1827, was the first suspension bridge built near London. It was replaced in 1887 by a new bridge. Hammersmith was formerly noted for its extensive market-gardens, orchards, and dairy farms. It had several good mansions, and was the residence of the nobility and wealthy citizens. Now the mansions have given place to factories and small houses, and all the fields have been built over. The Parish Church of St Paul was consecrated by Bishop Laud in 1631, and has been often repaired, enlarged, restored, and finally rebuilt. It has some interesting monuments; and among them is a bronze bust of Charles I. Hammersmith has quite an unusual number of large institutions of an ecclesiastical or educational character; besides St Paul's School itself, there are the Preparatory School (Colet Court), St Paul's Girls' School, Upper and Lower Latymer Schools, as well as Technical and County Council Schools. Queen Caroline, wife of George IV, died in 1821 at Brandenburg House, one of the most noteworthy of the Hammersmith mansions, and the house was soon afterwards pulled down. In recent years Hammersmith has been the home of many celebrated writers, among whom William Morris was one of the most famous.

Holborn is the smallest of all the London boroughs, and its population shows a density of 122 to the acre. The borough consists of the united parishes of St Giles-in-the-Fields and St George, Bloomsbury, St Andrew, and St George the Martyr, the Liberty of Saffron Hill, Lincoln's Inn, Gray's Inn, Staple Inn, and part of Furnival's Inn. The borough is divided into nine wards, and the council consists of seven aldermen and 42 councillors. Holborn plays an important part in the history of London, and its main road formed the route from Newgate and the Tower to the gallows at Tyburn. William, Lord Russell, went to the scaffold in Lincoln's Inn Fields, and Titus Oates and others were whipped in the Holborn line of road from Aldgate to Tyburn. Gerard dates his *Herbal*, 1597, from his house in Holborn, and in this famous book he mentions many of the rarer plants which grew well in his garden. Perhaps the greatest improvement in Holborn of recent years was the construction of the Holborn Viaduct, a really fine engineering feat, which spans Farringdon Street. It was opened in 1867 and has proved a great boon to the City, for now the traffic has no longer to climb Snow Hill or Holborn Hill.

Kensington is known as the Royal borough, and as the "Old Court suburb." It was specially provided by the Act of 1899 that Kensington Palace should be transferred from Westminster to Kensington. Among the London boroughs Kensington ranks twelfth in point of area, and ninth as regards population, the density of which is 75 to the acre. A part of Kensington Gardens is within the Royal Borough, and on its eastern boundary is Hyde Park. The borough is divided into nine wards, and the borough council consists of 10 aldermen and 60 councillors. The Town Hall is one of the finest in London. Kensington Palace, to which reference has been made, is mainly the work of Wren. Kensington is famous for its public museums and institutions, and fine houses. Within its borders are the

Victoria and Albert Museum, the Natural History Museum, the Imperial Institute and University of London, the Royal College of Music, and the Guilds' Technical College. Holland House is the stateliest piece of Jacobean architecture in London, and Leighton House with the Arab Hall is now maintained by trustees for the encouragement of the fine arts. Among the many men of note who have lived in South Kensington may be mentioned Macaulay, Thackeray, John Leech, Leigh Hunt, and Lord Leighton.

Lambeth is fifth in point of area, and third as regards population among the London boroughs. As much land remains to be built over, the density of population is only 73 persons to the acre. There are about 208 acres of open spaces in Lambeth, including Brockwell Park, Kennington Park, Vauxhall Park, Archbishop's Park, Ruskin Park, and Myatt's Fields. The borough has nine wards, and the borough council consists of 10 aldermen and 60 councillors. The fine municipal buildings are at Brixton, and the new London County Council Hall is being built on land beside the river in Lambeth. Besides various manufactures, such as soap and chemicals, Lambeth is specially celebrated for its potteries, which have existed for upwards of 200 years. Lambeth Palace is the chief historical building in the borough, and for centuries it has been the official residence of the Archbishops of Canterbury. St Thomas's Hospital is a fine group of buildings at the Westminster Bridge end of the Embankment, and opposite the Houses of Parliament.

Paddington is much smaller than St Marylebone, but has a larger population, the density being 105 persons to the acre. At the beginning of the nineteenth century it was described as "a village situated on the Edgware Road, about a mile from London." It made a great advance towards the middle of that century when the terminus of the Great Western Railway was built. It has 132 acres of open spaces, of which 102 acres are in

Kensington Gardens. The borough has nine wards, and the borough council consists of 10 aldermen and 60 councillors. The chief highway of Paddington is Harrow Road, a broad and winding thoroughfare. Near its eastern end is Paddington Green with a fine statue to Mrs Siddons, the celebrated actress, who was a resident in Paddington. The Paddington branch of the Grand Junction Canal runs through the borough and joins the main canal at Uxbridge more than 13 miles away. In the Bayswater Road is the beautiful Chapel of the Ascension, whose interior is decorated with mural paintings of Scriptural scenes, especially those associated with the Ascension. Behind this Chapel is the disused burial ground of St George's, Hanover Square. Here is the grave of Laurence Sterne, with the inscription on the tombstone, "Alas, poor Yorick! Near to this place lies the body of the Rev. Laurence Sterne, M.A., who dyed September 13, 1768, aged 55 years."

St Marylebone is one of the smaller London boroughs, and its population shows a density of 80 persons to the acre. The population and the number of inhabited houses show a steady decrease, owing no doubt to the conversion of dwelling houses into shops, factories, etc. This borough is fortunate in having 372 acres of open spaces, including 362 acres of Regent's Park. There are nine wards in the borough, and the borough council consists of 10 aldermen and 60 councillors. The name of this borough has undergone many changes. Pepys curiously renders it Marrowbone, and in the eighteenth century it was Marybone. Originally, however, the parish was called Tyburn, from the stream which ran through it. But at the beginning of the fifteenth century a new church dedicated to St Mary was built, and this was called St Mary's-le-bourne, i.e. St Mary on the brook, to distinguish it from other churches of the same dedication. We are told that the people of the village were glad to change its name from Tyburn, which was associated with the gallows, to that of the new church. Among the eminent residents in Marylebone have been James Martineau, the Brownings, the Hallams, and D. G. Rossetti.

St Pancras, the ninth of the London boroughs in point of size, and the seventh as regards its population, has a density of 80 persons to the acre. It is a parish of great antiquity and is named after St Pancras, a young Phrygian who suffered martyrdom at Rome. The borough has about 350 acres of open spaces including Parliament Hill, Waterlow Park, and part of Regent's Park. There are eight wards in the borough, and the borough council consists of 10 aldermen and 60 councillors. The Foundling Hospital, founded in 1739 by Thomas Coram, is a unique institution in this borough. University College and University College Hospital are two of the finest public buildings in St Pancras, while three of the great railway termini—St Pancras station, Euston station, and King's Cross station are within its borders. Among the noted residents in this borough have been Shelley, Dickens, Charles Dibdin, and Ruskin.

Wandsworth, which owes its name to the little river Wandle, is the largest of all the London boroughs. In point of population it ranks second, while the density is very low, being only 34 persons to the acre. The borough, consisting of the parishes of Clapham, Putney, Streatham, Tooting Graveney, and Wandsworth, is divided into nine wards. The borough council has 10 aldermen and 60 councillors. Wandsworth is fortunate in having no less than 1163 acres of open spaces, including Tooting Common, Putney Heath, Streatham Common, and parts of Wandsworth and Clapham Commons, and of Richmond Park. Wandsworth has a large number of public institutions, such as schools and hospitals, which have no doubt been attracted to it by the fine open spaces. Among the eminent men who have lived within its borders may be mentioned Thomas Cromwell, Cardinal Wolsey, Gibbon, the historian of Rome, and

Swinburne, the poet. Clapham is more particularly associated with the "Clapham sect," which included such names as Zachary Macaulay, William Wilberforce, and Granville Sharp. Wandsworth became a seat of several important manufactures introduced by the French Huguenots who took refuge in London after the Revocation of the Edict of Nantes. In 1748, Kalm, a Swedish naturalist, visited England, and his reference to Wandsworth is interesting. He says, "On the other side of the Thames opposite Fulham there lay a large and tolerably flat and bare common, which was abandoned to pastures. It was for the most part overgrown with *Genista spinosa*, furze, which was now in its best flower, so that the whole common shone quite yellow with it. In some places we saw ling here: but it was quite small. I also saw plats with Reindeer-moss (*Lichen rangiferinus*), which also was very short."

A Table giving the Area and Population of the City of Westminster, and the Boroughs in the North-west and South-west of the County of London.

The City of Westminster, and Boroughs	Area in Acres	Population in 1911
City of Westminster	2502·7	160,277
Battersea	2160·3	167,793
Chelsea	659·6	66,404
Fulham	1703·5	153,325
Hammersmith	2286·3	121,603
Hampstead	2265·0	85,510
Holborn	405·1	49,336
Kensington	2291·1	172,402
Lambeth	4080·4	298,126
Paddington	1356·1	142,576
St Marylebone	1472·8	118,221
St Pancras	2694·4	218,453
Wandsworth	9108·0	311,402
	32985·3	2,065,428

Note. The Administrative County of London, including the City of London, had a total area of 74,816 acres, and a population of 4,522,961 at the census of 1911.

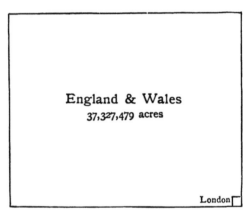

Fig. 1. Area of the Administrative County of London
(74,839 acres) compared with the area of England and Wales

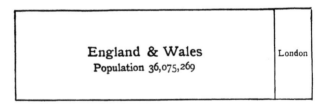

Fig. 2. The Population of the Administrative County of
London (4,522,961) compared with that of England and
Wales in 1911

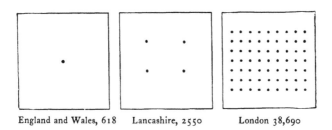

England and Wales, 618   Lancashire, 2550   London 38,690

Fig. 3. Comparative Density of Population to sq. mile
in 1911

(*Each dot represents 618 persons*)

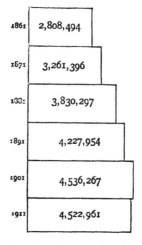

1861	2,808,494
1871	3,261,396
1881	3,830,297
1891	4,227,954
1901	4,536,267
1911	4,522,961

Fig. 4. The Growth of Population in London from
1861—1911

17—3

INDEX

Milton Keynes UK
Ingram Content Group UK Ltd.
UKHW032321161024
449665UK00001B/4

9 781107 663602